English Electric Type 2 Bo-Bo 'Baby Deltic' Locomotives

D5909 and D5902, Hitchin s.p., 18 August 1969. (Peter Foster)

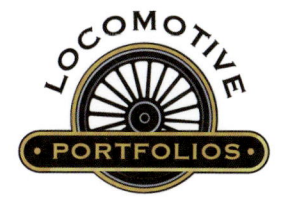

English Electric Type 2 Bo-Bo 'Baby Deltic' Locomotives

BR Type 23, Design to Destruction

ANTHONY P. SAYER

AN IMPRINT OF PEN & SWORD BOOKS LTD.
YORKSHIRE – PHILADELPHIA

First published in Great Britain in 2025 by
Pen and Sword Transport
An imprint of
Pen & Sword Books Ltd.
Yorkshire - Philadelphia

Copyright © Anthony P. Sayer, 2025

ISBN 978 1 39905 845 2

The right of Anthony P. Sayer to be identified as author of this work has been asserted by him in accordance with the Copyright, Designs and Patents Act 1988.

A CIP catalogue record for this book is available from the British Library.

All rights reserved. No part of this book may be reproduced or transmitted in any form or by any means, electronic or mechanical including photocopying, recording or by any information storage and retrieval system, without permission from the Publisher in writing.

Typeset in Palatino by SJmagic DESIGN SERVICES, India.

Printed and bound by Printworks Global Ltd, London/Hong Kong.

Pen & Sword Books Ltd. incorporates the imprints of Pen & Sword Books: After the Battle, Archaeology, Atlas, Aviation, Battleground, Discovery, Family History, History, Maritime, Military, Politics, Select, Transport, True Crime, Fiction, Frontline Books, Leo Cooper, Praetorian Press, Seaforth Publishing, Wharncliffe and White Owl.

For a complete list of Pen & Sword titles please contact

PEN & SWORD BOOKS LIMITED
George House, Beevor Street, Off Pontefract Road, Hoyle Mill, Barnsley, South Yorkshire, England, S71 1HN.
E-mail: enquiries@pen-and-sword.co.uk
Website: www.pen-and-sword.co.uk

or

PEN AND SWORD BOOKS
1950 Lawrence Rd, Havertown, PA 19083, USA
E-mail: uspen-and-sword@casematepublishers.com
Website: www.penandswordbooks.com

CONTENTS

Preface 6
Acknowledgements 8
Abbreviations 9

Chapter 1	Class Introduction 10
Chapter 2	Financial Considerations 17
Chapter 3	Technical Aspects 18
Chapter 4	Appearance Design and Styling 43
Chapter 5	Delivery and Acceptance 44
Chapter 6	Allocation History 54
Chapter 7	Major Overhauls and Repairs 62
Chapter 8	Locomotive Histories 66
Chapter 9	Performance and Service Problems (1959–63) 101
Chapter 10	Rehabilitation 127
Chapter 11	Re-Delivery and Acceptance 145
Chapter 12	Performance and Service Problems (1964–71) 147
Chapter 13	Mileages 155
Chapter 14	Accident and Fire Damage 158
Chapter 15	Operations: A High-Level Summary 160
Chapter 16	Details and Differences 177
Chapter 17	Liveries 185
Chapter 18	Storage and Withdrawal 191
Chapter 19	Storage Locations 198
Chapter 20	Disposal 203
Chapter 21	Departmental Service 216
Chapter 22	Preservation 219
Chapter 23	Concluding Remarks 220

Sources & References 223

PREFACE

The ten English Electric (EE) Type 2 locomotives were originally described collectively as the 'Small Deltics' reflecting their diminutive size compared with the prototype 'DELTIC' locomotive. However, this description very rapidly, and certainly by mid-1959, had morphed into 'Baby Deltics'.

As far as I am aware there has never been a book published devoted exclusively to the development, operation and demise of the 'Baby Deltics', the nearest being Brian Webb's outstanding book *The Deltic Locomotives of British Rail* (1982) which covered all of the 'Deltic'-engined locomotives which operated on British Railways. Various pictorial books have also provided 'full' 'Deltic' coverage (i.e. including the twenty-two Type 5 production 'Deltics') and the 'Baby Deltics' also shared their presence in a magazine 'part work' in 2017 with the 'Metrovick' Class 28s, a strange combination indeed! In contrast, however, there have been numerous magazine articles, which by their very nature were somewhat superficial.

This book, like its predecessors in the Pen & Sword 'Locomotive Portfolios' series, deliberately aims to introduce new insights to enable the further development of our knowledge of the 'Baby Deltic' fleet. This has been possible by the inclusion of material derived from:

(a) archive sources covering the background to the ordering of the class by the British Transport Commission, together with details surrounding the decision not to expand the class beyond the original ten 'Pilot Scheme' locomotives.

(b) extensive quantities of internal EE archive material (inter-departmental memos and correspondence with the BTC) regarding the mechanical trials and tribulations of the class which provide an in-depth understanding of the problems experienced by the 'Baby Deltics' and the remedial actions being pursued, including the whole process of rehabilitation during the 1963-65 period and the ultimately aborted re-engining of D5901 (or DP3 as it was intended to become).

(c) Diesel Locomotive Record Cards (DLRCs) for six of the ten locomotives, together with British Railways (Eastern Region, King's Cross District) locomotive shopping records.

(d) Motive Power Committee and CM&EE Technical archive reports which provide a quantitative as well as qualitative insight into 'Baby Deltic' performance, particularly with respect to mileages covered and their availability for traffic.

(e) extensive personal sighting data from numerous sources which has greatly assisted with broadening the information available on the class, particularly with respect to Works visits in the period post-rehabilitation. This information has proved instrumental in de-bunking much previously published information regarding locomotive storage and disposal.

Hopefully all this information will facilitate a greater understanding the Class 23 locomotives as the 'Baby Deltics' became under the

TOPS system. Individual histories of each of the ten locomotives are included. Availability of official and personal sighting information regarding works visits has enabled the locomotive livery histories to be substantially enhanced.

Space constraints have been less onerous not least because only ten locomotives are covered. This has given the opportunity to dig deep into the technical idiosyncrasies of the Napier 'Deltic' engines together with the context of why such sophisticated engines came to be deployed on mundane workaday Type 2 locomotives.

References are made in the text to No.1 and No.2 ends of locomotives. For clarity, No.1 end is defined as the end containing the radiator compartment. Mention is also made to '1-2' or '2-1' sides of the locomotives; the '1-2' side is the side of the locomotive where the No.1 end is to the left and No.2 end to the right.

Anthony Sayer
2024

D5908, Doncaster Works, 31 May 1959. (Howard Forster [David Dunn Collection])

ACKNOWLEDGEMENTS

Archive sources have provided a major input in developing the class and individual locomotive histories and I would like to personally thank the teams of people at The National Archives (Kew), the National Railway Museum (York) and the Essex Record Office (Chelmsford) for their kind assistance.

Inputs from David Percival, Dave Brennand and Graham Hardinge have collectively provided a considerable amount of local information and context. The Stratford DRS Arthur Nugent and Aubrey Rayment reports also proved to be particularly useful.

Considerable inputs from both Mike Hunt and Mark Parsons have helped substantially with fully understanding the complexities of the Napier 'Deltic' engine, the extensive array of English Electric electrical equipment deployed and the associated auxiliary paraphernalia. It was over fifty years ago that I took my 'A' level Physics examination so some of this was somewhat outside of my comfort zone but Mike and Mark managed to successfully guide me through. Thank you both.

Thanks are also due to Fred Kerr in pointing me in the direction of additional English Electric archive material beyond that which I was already aware. And very useful it proved to be! Assistance with engineering drawings was very generously supplied by Grahame Wareham.

My continued use of the 'back to basics' approach relied heavily on a significant amount of personal observations in various forms. I have made every attempt to credit everyone concerned in the References and Sources section. As before, my thanks go to you all.

This book has used approximately 210 images to support the text. My sincere thanks go to the following organisations for the photographs used herein:

Paul Chancellor (Colour-Rail), Paul Appleby (Rail Photo Archives), John Chalcraft (Rail-Photoprints), David Bird (RCTS Archive), Robin Fell (Transport Treasury), Michael Mercer (Rail Image Collections), Rail-Online and Old-Maps.co.uk.

Also my appreciation goes to the following individuals:

Bill Atkinson, Jim Binnie, Stewart Blencowe, Adrian Booth, Tom Bowman (via 'Griffith_p'), Chris Burton, J.R. (Jim) Carter (Footplate Cameraman), Jim Carter ('Thru the Round Window' Collection), Graham Clark (via Alan Monk), Stephen Dowle, David Dunn (Collection), John Evans, Peter Foster, Peter Groom, Graham Hardinge, Keith Holt, Peter Ingarfill (via Peter Foster), Fred Kerr, Gordon Lacy (via Nigel Lacy and "Chronicles of Napier"), John Law, Brian Lee, Colin Marsden (Collection), Alistair Ness, Roger Norfolk, David Percival, Nigel Petre, Jack Ray (via Stuart Ray), Geoff Sharpe, R.H.G. Simpson, Tony Skinner, Steve Thorpe, John Grey Turner, Charlie Verrall, Grahame Wareham, Alan Whincup and George Woods.

There are images from a very small number of photographers where it has proved impossible to obtain the appropriate permission to use their photographs despite every endeavour to do so. In addition, there are one or two images where the identity of the photographer is totally unknown; lack of accreditation on slides or prints has prevented any possibility of determining their provenance. In both cases, anyone who feels that they have not been adequately credited should please contact me via the publisher to ensure that the situation is corrected in future editions.

ABBREVIATIONS

General Abbreviations

AEI	Associated Electrical Industries.
BLS	British Locomotive Society.
BR	British Railways/British Rail.
BRB	British Railways Board.
BRB SC	BRB Supply Committee.
BRB W&EC	BRB Works & Equipment Committee.
BRCW	Birmingham Railway Carriage & Wagon Co. Ltd.
BTC	British Transport Commission.
CM&EE	Chief Mechanical & Electrical Engineer.
CS or c.s.	Carriage Sidings.
DLRC	Diesel Locomotive Record Card.
DRS	Diesel Repair Shop (Stratford).
ECS or e.c.s.	Empty coaching stock.
EE	English Electric.
ER	Eastern Region.
FDTL	*"Fires on Diesel Train Locomotives"* Reports.
GE	Great Eastern.
GN	Great Northern.
LCGB	Locomotive Club of Great Britain.
MPD	Motive Power Depot.
MV or Metrovick	Metropolitan Vickers
NBL	North British Locomotive Co. Ltd.
NTP	National Traction Plan.
P/e or p/e	Period Ending.
RCTS	Railway Correspondence and Travel Society.
rpm	Revolutions per minute.
RSL	Rolling Stock Library.
RTC	Railway Technical Centre, Derby.
SLS	Stephenson Locomotive Society.
s.p. or sp	Stabling point.
S(u)	Stored unserviceable.
TOPS	Total Operations Processing System.
T&RS	Traction & Rolling Stock.
W/e, or we	Week ending.

English Electric Abbreviations

RSDD	Rolling Stock Design Department (Preston).
TCDD	Traction Control Design Department (Bradford).
TS&C	Traction Sales & Contracts (Bradford).
VF	Vulcan Foundry, or, Vulcan Works (Newton-le-Willows).

Chapter 1
CLASS INTRODUCTION

1.1 Background

The English Electric Company Ltd was formed in 1918 from the merger of Dick, Kerr, and Co. Ltd, Preston, and, Phoenix Dynamo and Manufacturing Co. Ltd, Bradford, the former having already owned Willans & Robinson Ltd of Rugby since 1916. It also purchased the works of Siemens Bros. Ltd, Stafford.

An association between R&W Hawthorn, Leslie & Co. Ltd, Newcastle-upon-Tyne and English Electric started in the 1920s when Hawthorne Leslie built a number of electric locomotive mechanical parts for EE for use in locomotives being erected in Preston. In 1933, EE decided to build its first diesel-electric locomotive and the mechanical construction was entrusted to Hawthorn Leslie.

In 1937 Hawthorn Leslie amalgamated with Robert Stephenson & Co. Ltd, Darlington, to become Robert Stephenson & Hawthorn Ltd (RSH), with RSH taking over the locomotive building department of Hawthorn Leslie at its Darlington facility.

English Electric erected locos in Preston and subcontracted work to Vulcan Foundry Ltd (VF), Newton-le-Willows; this arrangement ultimately resulted in the cessation of erecting work at Preston.

In 1942, English Electric took over D. Napier & Son, an aero-engine manufacturer, which, under EE, expanded activities to include marine and railway applications.

RSH and VF formally merged in 1944, and in March 1955 they both merged into English Electric, with the manufacture of EE locomotives then being carried out at both Darlington and Newton-le-Willows. In due course, RSH and VF were fully integrated into the EE Group as the Vulcan and Stephenson works. In terms of locomotive manufacture English Electric were now in a position to provide a fully integrated operation manufacturing engines and electrical gear, together with erection capability.

A decline in orders brought closure of the Darlington works in 1964 with all locomotive construction being concentrated at Newton-le-Willows.

In 1967, English Electric-AEI Traction was formed as a consequence of the merger of English Electric and Associated Electrical Industries (AEI), which already incorporated British Thompson, Houston Ltd and Metropolitan Vickers Ltd, and, in combination constituted the largest and arguably the most experienced rail traction group in the world.

An even larger grouping followed in 1968 when General Electric Co. Ltd (GEC) took over the whole of the EE-AEI group. Severe rationalisation followed which culminated in the elimination of locomotive manufacture at Vulcan Works by 1969/70, with spare capacity in other manufacturers' works in the GEC group deployed instead.

1.2 'Baby Deltic' Order Placement

The BTC Works & Equipment Committee (W&EC) Meeting Minute 293 (dated 17 November 1954) recorded approval in principle for 160 diesel-electric locomotives at a total estimated cost of £9.28m, 130 for construction by contractors and 30 in BR Workshops.

The breakdown of the locomotives to be built by contractors was forty in the 600-800hp range (Type A), eighty in the 1,000-1,250hp range (Type B) and ten in the 2000-and-over hp range (Type C); locomotives to be built by BR Workshops were split twenty in the 1,000-1,250hp range and ten in the 2000-and-over hp range.

The remaining fourteen locomotives of the so-called 'Pilot Scheme' fleet of 174 locomotives were constructed with hydraulic transmissions.

The twenty diesel-electric locomotives in the 2000-and-over hp range were approved by the W&EC on 9 February 1955 (Minute 357/40) and authorised by the BTC on 17 February 1955 (Minute 8/74[o]) at a cost of £1.68m and were included in the 1956 Building Programme. The W&EC Meeting Minute 388/26 (dated 23 March 1955) approved the construction of the remaining 140 diesel-electric locomotives provisionally as part of the 1957 and 1958 Locomotive Building Programmes, at a total estimated cost of £7.60m. Advance authority was given by the BTC on 24 March 1955 (Minute 8/142[i]).

Tenders for the 130 complete locomotives and power equipment for the 30 locomotives to be built in BR Workshops were invited from sixteen contractors, including five overseas companies. Notwithstanding gauge considerations, the tenders received from firms abroad were higher than for comparable British types of locomotives and, therefore, only tenders submitted by the selected British companies were accepted.

The recommended tenders (i.e. those which were considered to best meet requirements) for the 130 complete diesel-electric locomotives to be built by contractors amounted to £8,513,450 and for the thirty power equipment sets for the locomotives to be built in British Workshops, £1,478,600 i.e. a total of £9,992,050 against the originally authorised expenditure of £9,280,000, an increase of £712,050 (7.6%).

With respect to the locomotives in the 1,000-1,250hp category, five contractors were nominated for the supply of the required eighty locomotives, as follows:

Main Contractor	No. of Locos	Total Price	Delivery Period
English Electric	10	£ 714,000	30 months
Metropolitan-Vickers	20	£1,428,000	24 months
Birmingham Railway Carriage & Wagon	20	£1,260,750	20-23 months
Brush Electrical Engineering	20	£1,413,000	21 months
North British Locomotive Co.	10	£ 624,000	21 months
		£5,439,750	

It will be noted that a strict least-cost selection process was not applied, the aspect of lowest cost being conditioned by the wish to gain experience with a wide variety of makers and designs. However, the combination of engine builders and electrical contractors recommended was 'mainly governed by the selection of the lowest combined tender' where alternatives were available.

A memorandum dated 29 September 1955 submitted by Chief Electrical Engineer, Chief Mechanical Engineer, Chief Financial Officer, Chief Operating and Motive Power Officer, Chief Commercial Officer and Chief Stores Officer to the W&EC recommended acceptance of the tenders for the 160 locomotives at a total cost of £9.992m and the proposed distribution of orders between manufacturers. The W&EC approved the memorandum on 12 October 1955 (Minute 517/16).

An order for ten Type B Bo-Bos was placed by the BTC with English Electric on 16 November 1955. The 1957 Locomotive Building and Condemnation Programmes were presented to the W&EC, via a Memorandum dated 2 December 1955, ready for their meeting on 7 December 1955; W&EC Minute 567 (Supplementary Item No.30) recorded the inclusion of the ten English Electric locomotives and the Building Programme was accepted and submitted to the BTC for authorisation which was duly completed on 15 December 1955 (Minute 8/590).

The excess tender prices for the Type A and B 'Pilot Scheme' locomotives over the original March 1955 estimates were authorised as part of the 1957 Building and Condemnation Plans, hidden away amongst other orders for diesel shunters and steam locomotives and residual scrap values for redundant steam locomotives.

An order for the ten Type B locomotives was placed by English Electric on Vulcan Foundry on 23 January 1957 for the mechanical portion of the 'Baby Deltics' and locomotive erection under Contract No. CCF0875.

D5900, BR Doncaster Works Official Photograph, 1959.
(Grahame Wareham Collection)

1.3 Why the High-Speed Deltic Engine?

At the time when the BTC were inviting the various locomotive manufacturers to offer prospective designs against the pre-defined Pilot Scheme power categories, English Electric did not have a conventional engine which naturally fitted the Type B category. Their 8-cylinder engine design as offered in the Type A locomotives (subsequently D8000-19) was not yet available with charge air cooling and was, therefore, restricted to 1,000hp maximum, and the 12-cylinder engine was too heavy to be accommodated in a locomotive defined as having a maximum weight of 72tons, or, more specifically, 18tons axle-loading.

D. Napier, a subsidiary of English Electric, were looking to widen the market their 'Deltic' engines beyond maritime applications and saw railway locomotives as an attractive additional outlet for their 9- and 18-cylinder engines. English Electric had already deployed the Napier 'Deltic' engine in the 'DELTIC' prototype. The turbocharged T9 unit, rated at 1,100hp, represented an ideal solution for the BTC Type B power category.

1.4 Drawing the Line at Ten

The 'Pilot Scheme' provided six Type B (later Type 2) designs for testing purposes over a three-year period (five from contractors and one from BR Workshops); this quite deliberate objective was intended to ultimately determine a limited number of types for volume production commencing 1961/62 based on operational experience gained.

However, facing deteriorating financial results, the BTC abandoned the three year trial period, and extended and accelerated the introduction of diesel locomotives as fast as British production capacity would allow, believing that dieselisation and the consequent elimination of steam would dramatically improve the position. Minute 9/384 of the BTC Meeting of 26 July 1956 recorded that:

The Commission...... discussed the purchase of additional main line locomotives and agreed that they would be prepared to consider requests for a number of these without further trials, provided that:

1. There is sufficient technical evidence to show that the type of locomotive desired is fully and without doubt able to meet requirements which are comparable to those in the service for which it is intended.
2. The substitution of the diesel locomotives for steam locomotives is economically justified by the manner in which they will be operated.

The dangers of giving up the trial period were made clear by R.C. Bond (Chief Mechanical Engineer) but Sir Brian Robertson (BTC Chairman) insisted that the Board's decision was adhered to with the specific condition that good reliable locomotives were introduced.

The Commission also insisted that the number of different designs of locomotives should be reduced to an absolute minimum. This, as already mentioned, was one of the key objectives of the 'Pilot Scheme' process but it had now become necessary to recommend the smallest possible number of types without any operating experience having been obtained with the locomotives then on order. The only way of achieving this was to base recommendations on engineering judgement, knowledge of various firms' products, and the operating experience of other railways.

A Memorandum to the Works and Equipment Committee dated 20 September 1956 proposed, under the requirements of BTC Min. 9/384, that fifty-six Type B (1,000-1,250hp) locomotives should be included in the 1958 Locomotive Building Programme i.e. ten to be built in BR Workshops at an estimated cost of £760,000 (D5020-9) and forty-six by contractors at an estimated cost of £3,496,000. This proposal was approved by the W&EC on 26 September 1956 (Minute 750 [Item 26]) and passed on to the Commission for final authorisation. Minute 9/477 of the BTC meeting of 27 September 1956 recorded the necessary authorisation.

A Memorandum to W&EC dated 15 November 1956 proposed that five contractors should be invited to tender for the forty-six locomotives, these being Brush, English Electric, Metropolitan-Vickers, North British Locomotive (NBL), and Birmingham Railway Carriage & Wagon (BRCW). This list reflected the desire not to proliferate types beyond those already part of the 'Pilot Scheme' types. The W&EC approved the shortlist at their meeting on 21 November 1956 (Minute 782 [Item 4]).

Following receipt and analysis of the tenders, a further Memorandum to the W&EC dated 4 April 1957 recommended the placing of orders for the forty-six locomotives with NBL (twenty-eight [D6110-37] and BRCW [eighteen (D5320-D5337]), at a total cost of £3,162,080. Minute 876 (Item No.15) of the W&EC meeting of 10 April 1957 recorded approval for the placing of these contracts. Clearly English Electric were unsuccessful with their bid, but it is unknown if this was based purely on price or whether other issues were also considered. The remaining ten locomotives of the 1958 Building Programme (D5020-D5029) were allocated to be constructed in BR Workshops and deployed Sulzer engines.

A Memorandum to the BTC dated 16 May 1957 acted as a cover note for a Report entitled 'Modernisation of British Railways: Report on Diesel and Electric Traction and the Passenger Services of the Future'. This report was described as 'the definitive modernisation plan for traction and passenger traffic' and effectively updated significant parts of the 1955 Modernisation Plan. It highlighted 'the very much more rapid introduction of diesel locomotive traction than had first been intended', envisaging the ordering of a further 1,889 main-line diesel locomotives during the period 1957-62.

The BTC at their meeting on Meeting on 23 May 1957 (Minute 10/212) approved the general concept of the extension and acceleration of introduction of diesel traction as contained in the report, without commitment to the exact pace of extension, stating that:

> The Commission would be prepared to go further than they have already gone in regard to ordering diesel main line locomotives, in spite of the risk of unsatisfactory performance in the early stages, if the Regions presented them with a limited number of firm plans for their use in specific areas, containing as clear a justification as possible.

The Commission estimated that 'The maximum number of diesel main line locomotives likely to be procurable was about 500 in one year, and the Commission would be prepared to include up to this number in their locomotive orders for 1959 to meet the requirements of such schemes.'

The Commission asked that the question of limiting the number of types of main-line diesel locomotives types be specifically addressed, and to advise on what was practicable. The Report 'Main Line Diesel Locomotives: Limitation of Variety' (R.C. Bond & S. Warder, 26 July 1957) was produced in response to this request.

At the time of this report, 444 locomotives had been authorised, of which orders had been placed for 230 locomotives or equipment sets i.e. the original 174 Pilot Scheme Locomotives authorised in the 1956 and 1957 Building Programmes and 56 Type 2 locomotives authorised for the 1958 Programme. The remaining 214 authorised locomotives included 116 diesel-hydraulic locomotives for the Western Region and 98 diesel-electrics for the Southern Region.

The two key principles used in the 'Limitation of Variety' report governing the selection of locomotive types for ordering beyond those already ordered were reliability and as much standardisation as British production capacity would allow. On the basis of these key considerations, the recommendation was that any diesel-electric locomotive orders placed in the 1959 Building Programme should be limited to the following types (now classified as Types 1 to 4, an expansion of the previous A to C), subject to the phasing of the Regions' specific requirements and available manufacturing capacity:

Locomotive Type	Engine	Transmission	Mechanical parts
Type 1 1,000h.p.	English Electric	English Electric	English Electric
Type 2 1,160h.p.	Sulzer	BTH	BR
Type 3 1,750h.p.	English Electric	English Electric	English Electric
Type 4 2,500h.p.	Sulzer	Crompton	BR

English Electric and Sulzer 'conventional' engines featured strongly. It was considered that EE had the largest experience and productive capacity of any British manufacturer and had the resources to ensure the delivery of a reliable product, whilst the Sulzer engine was the best known and widely used outside of the USA and was recognised for excellent design and workmanship. Paxman engines were proposed as a reserve type, whilst Crossley and Mirrlees engines were not included in the recommendations list on the basis that both companies were deemed to have less experience of traction requirements than either EE or Sulzer.

Subsequent discussions with industry soon showed, however, that it was not possible to adhere strictly to the recommendations. To meet BR's heavy demand for Type 2 locomotives and to capitalise on available production capacity, Brush, BRCW and NBL were subsequently awarded orders for more of their Type 2 diesel-electric locomotives.

In reaction to the BTC decision recorded in Meeting Minute 10/212 (23 May 1957), the Regions submitted a range of Area Schemes requiring a total of 782 diesel locomotives, significantly in excess of the suggested 500 for construction during 1959. However, in a Memorandum to the BTC dated 11 September 1957, the W&EC suggested that the Commission approve a 1959 Building Programme of circa 782 diesel locomotives and at the BTC Meeting on 19 September 1957 (Minute 10/400), the Commission approved, in principle, a 1959 Building Programme of the order of 750 to 800 locomotives composed of the types recommended in the 'Limitation of Variety' report. It was recognised that this was a large requirement and production capacities may dictate some deferral into 1960. However, a subsequent government-imposed capital investment restriction in late 1957 severely limited the purchase of locomotives for delivery in 1959 and it was, therefore, impossible to take full advantage of building capacity for main-line diesel locomotives in that year.

The revised 1959 Building Programme included eighty-four Type 2 diesel-electric locomotives (plus shunting and electric locomotives); all of these Type 2s were to be built in BR Workshops employing Sulzer engines (subsequently D5030-D5113). The W&EC approved the building of the eighty-four locomotives at their meeting on 22 January 1958 (Minute 1110, Item No.20) and they were authorised by the BTC on 13 February 1958 (Minute 11/53). The remaining diesel locomotive requirements were deferred into 1960 and 1961. Contactors were invited to tender in advance of these later Building Programmes, but exactly which contractors were involved is not known.

The financial situation eased during 1958 and as a consequence there were two Supplements to the 1959 Building Programme, with locomotive types selected on the basis of tenders already received, as follows:

First Supplement (ninety locomotives), including the following Type 2s:

- 40 Brush (D5520-D5559) Authorised 26 June 1958.
- 20 NBL (D6138-D6157) Authorised 10 July 1958.

Second Supplement (forty-nine locomotives), including twenty-nine Type 2s:

- 20 Brush (D5560-D5579) Authorised 20 November 1958.
- 9 BRCW (D5338-D5346) Authorised 20 November 1958.

None of the locomotives in the First and Second Supplements featured in the 'Limitation of Variety' recommendations but were repeat orders for locomotives of types already in service. The key question is….. was English Electric invited to tender for the 1960/61 Type 2 requirements? And, with competitive quotes, could they have been awarded orders?

In late 1958, a memorandum was produced for discussion at the Chairman's Conference on Modernisation; this was produced by R.C. Bond (by now BTC Technical Advisor) and entitled 'Main Line Diesel Locomotives: The Approach to Standardisation' (dated 10 November 1958). The recommendations concerning the preferred diesel-electric manufacturers for each of the Types 1-4, as detailed in the 'Limitation of Variety report', were reiterated. However, this memorandum went a stage further by explicitly listing those designs which it was proposed should be *excluded* from any future orders i.e.

Type	Manufacturer	No. Series	Engine	Electrical Equipment
Type 1	NBL	D8400	Paxman	General Electric Company
Type 2	Metropolitan-Vickers	D5700	Crossley	Metropolitan-Vickers
	English Electric	D5900	Napier 'Deltic'	English Electric
	NBL	D6100	M.A.N.	General Electric Company

Notes of the Chairman's Conference, which took place on 13 November 1958, record that 'The Chairman … fully agreed with the proposals set out in the Technical Advisor's paper for the elimination of a number of locomotive types.' Bond's memorandum also supported repeat orders for types of locomotives not included in the 'Limitation of Variety' recommendations i.e. BTH Type 1, Brush and BRCW Type 2s and English Electric Type 4, to enable financial capital allocations to be used to the full during 1959/60.

A further report covering limitation of variety was produced in 1959 entitled 'Standardisation of Main Line Diesel Locomotives' (BR General Staff, 8 June 1959). By this date approximately 150 diesel-electric main line locomotives were in service; however, there was still insufficient experience with the new traction to statistically challenge the logic of the original 'Limitation of Variety' report recommendations. This report, once again, explicitly listed the 'excluded' types of diesel-electric locomotives. Following on from this report and in a Memorandum to the BTC from the General Staff dated 22 June 1959, the Commission were recommended to:

(a) endorse the principles embodied in the Report on 'Limitation of Variety' of main line diesel locomotives (nearly two years after it was published in 26 July, 1957!),
(b) note the further progress currently attained towards standardization (*as described in the report*), and,
(c) approve the proposal to specifically exclude the four specified types from the forthcoming 1960 and 1961 programmes.

The BTC Minutes of the Meeting on 25 June 1959 (Minute 12/253) recorded that the Commission accepted these recommendations with only a few minor caveats. Quite why it took nearly two years to formally ratify the 'Limitation of Variety' report is unclear, although it has to be said that the 1957 report, whilst clear about what was included in the list of recommended types, was not explicit about which types should NOT be included.

This lack of clarity, against the backdrop of expanded and accelerated production pressures and combined with manufacturing constraints across the UK, allowed the NBL and Brush Type 2 diesel-electric fleets to be expanded from 10 to 58, and, from 20 to 80 respectively (and ultimately 263 in the latter case). Theoretically these pressures could also have allowed the English Electric Type 2 fleet to expand had their tender for at least some of the 56 Type 2 locomotives included in the 1958 Locomotive Building Programme been financially more attractive, but beyond that the door had been firmly closed.

D5901. Stratford DRS, 21 May 1960. By this time any thought of expanding the 'Baby Deltic' fleet beyond the ten 'Pilot Scheme' examples had been quashed. (Norman Browne [RCTS Archive])

1.5 Timeline

Key Milestones	Date	Comments
W&EC approval for 100 Type 2 'Pilot' diesel locos	23/03/55	
BTC advance authority for 'Pilot' locos	24/03/55	
Memorandum to W&EC re. composition of 'Pilot' fleet	29/09/55	
W&EC Approval for 'Pilot' fleet including 10 Type 2 locos from English Electric)	12/10/55	
BTC order placed for 10 locomotives from English Electric	16/11/55	
Formal BTC approval of final prices of 'Pilot' fleet	15/12/55	
BTC extension/acceleration of diesel orders (1) (Min.9/384)	26/07/56	
1958 Locomotive Building Programme authorised	27/09/56	
EE order placed on Vulcan Foundry for manufacture of the mechanical portion of the Type 2s and locomotive erection under Contract No. CCF0875	23/01/57	
Placement of orders against 1958 Building Programme	10/04/57	Inc. 56 Type 2s; no EE locomotives.
BTC extension/acceleration of diesel orders (2) (Min.10/212)	23/05/57	
R.C. Bond/S. Warder *Limitation of Variety* Report	26/07/57	MV, NBL, Brush and BRC&W not included in the recommended manufacturers list.
1959 Locomotive Building Programme authorised	13/02/58	84 Type 2s (all to be built in BR Workshops).
1959 Supplementary Locomotive Building Programmes authorised (to capitalise on available production capacity)	26/06/58+ 20/11/58	Including 89 Type 2s (60 Brush, 20 NBL, 9 BRCW).
R.C. Bond Report: *Main Line Diesel Locomotives: The Approach to Standardisation*	10/11/58	EE Type 2 explicitly excluded from future orders.
First EE Type 2 (D5903) into traffic	17/04/59	
Central Staff Report *Standardisation of Main Line Diesel Locomotives*	08/06/59	EE Type 2 explicitly excluded from future orders.
BTC acceptance of *Limitation of Variety* Report	25/06/59	

D5900, with NBL Type 2 D6114, EE Type 4 D212 and BRC&W Type 2 D5326, Doncaster Works (Paint Shop Yard), 22 May 1959. D5900 surrounded by three locomotive types which were ultimately proliferated beyond the 'Pilot Scheme' order quantities. All locomotives undergoing acceptance testing.
(Grahame Wareham Collection)

Chapter 2
FINANCIAL CONSIDERATIONS

Following advanced authority for the 'Pilot Scheme' locomotives by the BTC in March 1955, contractors were invited to tender for the various Types required. Design details and associated prices were assessed and in September 1955 the English Electric design with the 'Napier' engine was accepted as one of the designs to be included in the scheme at a price of £71,400 each, subject to price variation which allowed for inflation (covering wages, raw material costs, etc) between the time of order placement and the contracted locomotive delivery date.

The price actually paid was £79,050 each (based on accounting information quoted within 1968-71 Condemnation documentation).

D5903, King's Cross, 1960. (Grahame Wareham Collection)

Chapter 3
TECHNICAL ASPECTS

3.1 Overview

The 'Baby Deltic' locomotives were built under English Electric contract CCF0875. English Electric was the main contractor for these locomotives and provided the electrical equipment. The mechanical portion was built and locomotive erection undertaken by Vulcan Foundry Ltd, Newton-le-Willows, with the order placed on 23 January 1957; the VF-allocated order numbers were 6553 and 6554 of 1957, with the locomotives built in two batches of five. The engines were supplied by D. Napier & Son Ltd. Vulcan Foundry and Napier were both part of the English Electric group of companies.

Consistent with other 'Pilot Scheme' diesel locomotive contracts placed by BR at the time, a list of required standard equipment and preferred fittings was included. The intended duties for the 'Baby Deltics', dictated a maximum overall height 12ft 8in, a maximum weight of 72tons (18t axle-load) and the provision in the vacuum brake pipe circuit for the fitment of trip-cock gear; all of these requirements were based on the planned use of the 'Baby Deltics' on Great Northern suburban duties, including services on the Metropolitan Widened Lines to Moorgate, and cross-London inter-regional freight workings to the Southern Region.

Some of the details included in the Purchase Order placed on Vulcan Foundry by English Electric are given below:

The English Electric Co. Ltd, Preston
Purchase Order: PT.1032.V.7. 23/01/57.
VF No. 6553-61/57
Working No. 04CCF0875/41

British Transport Commission
10 Sets   Mechanical Parts for Type 'B' 1,100hp Bo-Bo diesel-electric locomotives including erection of the diesel-electric equipment in the locomotive.

VF Nos.
6553/57   5 Sets Mechanical Parts for 1,100hp Bo-Bo diesel-electric locomotive.
6554/57   5 Sets Mechanical Parts for 1,100hp Bo-Bo diesel-electric locomotive.
6555/77   Installation of electrical equipment for VF6553/57.
6556/77   Installation of electrical equipment for VF6554/57.

Details
Mechanical Parts for Type 'B' 1,100hp Bo-Bo Diesel-Electric locomotives with train heating boilers. The locomotives are to have two cabs and nose end compartments and are to be generally in accordance with Drawing P.3200/271.

Bogies are to be generally similar to those being manufactured by us for the BTC Type 'A' locomotives.

Independent fuel tanks are required for the engine and boiler, these are to have capacities of 450 gallons and 100 gallons respectively.

Water tanks having a total capacity of 600 gallons are required, and no water pick-up is to be provided.

Provision is to be made for the future fitting of the BTC-type Automatic Train Control apparatus.

Each set of mechanical parts is to be complete with:

(a) One Westinghouse 2EC72A mechanically-driven compressor to Westinghouse Drawing D.6327 Pt. No. D73954/7.
(b) Two Northey type 125RE exhausters.
(c) One train heating boiler of 1600lb/hr capacity to be

of the Vapor-Clarkson or Clayton type. Details will be settled later.

Division of Work
In accordance with usual arrangements, with the addition that we are to be responsible for obtaining and fitting the compressor and exhauster equipments, the train heating boiler, and for erection of the diesel-electric equipment in the locomotives.

Conditions of Contract
In accordance with BTC1.

Inspection & Test
In accordance with Clauses 15 and 16 of BTC Specification DE/M/1.

Leading particulars are set out in the following section and it is recognised that some details are at minor variance with the above purchase order. Construction commenced at Vulcan Foundry in late 1957.

3.2 Leading Particulars

Engine: Napier 'Deltic' T9-29, 9-cylinders, opposed-piston, two-stroke, turbo-charged, compression-ignition.
 5.125in. cylinder bore x 7.25in. piston stroke (two pistons per cylinder).
 Maximum continuous rated output: 1,100hp at 1600rpm.

Main Generator: English Electric EE835A (or EE835D – see Sections 3.12.1 and 10.3.4).
 6-pole, 400V, 1700A, 680kW, 1600rpm.

Auxiliary Generator: English Electric EE912/1B.
 4-pole, 110V, 410A, 45kW, 1,000/2,680rpm.
 Output (included in rated bhp): Up to 90bhp at 2680rpm.

Traction Motors: Four English Electric EE533A, six-pole, force-ventilated, nose-suspended motors; single-reduction gear drive (63:17 gear ratio).
 Continuous rating: 250V, 850A, 248hp, 350rpm; One hour rating: 250V, 900A, 261hp, 340rpm.

Traction Motor Blowers: Two Aerex Type 2TP040.

Vacuum Exhauster Motor: Two EE755/1A, changed to two EE762A prior to delivery.

Vacuum Exhauster: Two Northey Boyce RE125, changed to two Reavell (5.25inx10in) prior to delivery.

Air Compressor: Mechanically-driven Westinghouse Type 2EC72A, substituted by electrically-driven compressor at rehabilitation (see Section 10.3.8).

Fuel Transfer Pump: Varley DH25.

Fuel Transfer Pump Motor: EE TFZ97918.
 Circulating pressure: 20 lb/sq.in. (1.406 kg/cm²).

Injection Pumps: Napier/CAV type FM 110 B5, one per cylinder.

Injectors: CAV nozzle type BDL 15 6230, one per cylinder.

Load Regulator Motor: EE TFZ97904.

Batteries: D.P. Battery Co. Type RSKB 138M/6 (48 cells).

Radiator: Marston Excelsior double-bank radiator, Spiral Tube shutters.

Radiator Fan: Airscrew and Jigwood fan (engine-driven through gearbox). Fan diameter 54in, eight blades (subsequently increased to twelve).

Performance:
Maximum Tractive Effort: 47,000lb at 29.1% adhesion at 2400A main generator.

Continuous Tractive Effort: 30,600 at 9.4mph at 1700A main generator; also separately listed as 31,800lb at 10mph.

Rail hp at Continuous Rating: 768hp.

Full engine output: Available between 7 and 75mph.

Braking: Air for locomotive, vacuum for train giving a brake force of 82.5% of locomotive weight in working order.

Train Heating Equipment: Stone Vapor OK4616 boiler, steaming capacity 1600lb/hr.

Tank Capacities:
Engine and Boiler Fuel Tank: 550gal.

Boiler Water Tank: 500gal.

Overall Length: 52ft 6in.

Overall Height: 12ft 6in. (over superstructure framing).
Overall Height: 12ft 8in. (over covers and hatches).

Overall Width: 8ft 10¾in.

Wheel Diameter: 3ft 7in.
Bogie Wheelbase: 8ft 6in.
Total Wheelbase: 40ft 6in.

Distance between Bogie Pivot Centres: 32ft 0in.

Total Weight:
73t 17cwt (in working order).
69t 0cwt (empty/dry).
15t 0cwt (unsprung).

Component Weights:
Engine (dry): 3.33tons.

Main Generator: 3.7tons.
Bogie (complete): 13.2tons.
Train-Heating Boiler: 1.67tons.

Wheel Arrangement: Bo-Bo.

Axle Weights:
No.1-end bogie (outer axle): 18t 10cwt
No.1-end bogie (inner axle): 18t 5cwt.
No.2-end bogie (inner axle): 18t 11cwt.
No.2-end bogie (outer axle): 18t 11cwt.

Maximum Speed: 75mph.

Minimum Curve negotiable: 4 chains minimum radius curve, without gauge widening, at dead slow speed.

Minimum Vertical Curve negotiable (Convex and Concave): 11 chains minimum radius curve.

Route Availability: 5.

An internal EE memo dated 8 January 1959 provided the additional weight information for the following key items of equipment:

One traction motor	5600lbs
One radiator panel (dry)	1022lbs
One exhauster	1218lbs (presumably the original Northey Boyce exhauster)
Train Heating Boiler (dry)	3465lbs

3.3 Diagrams

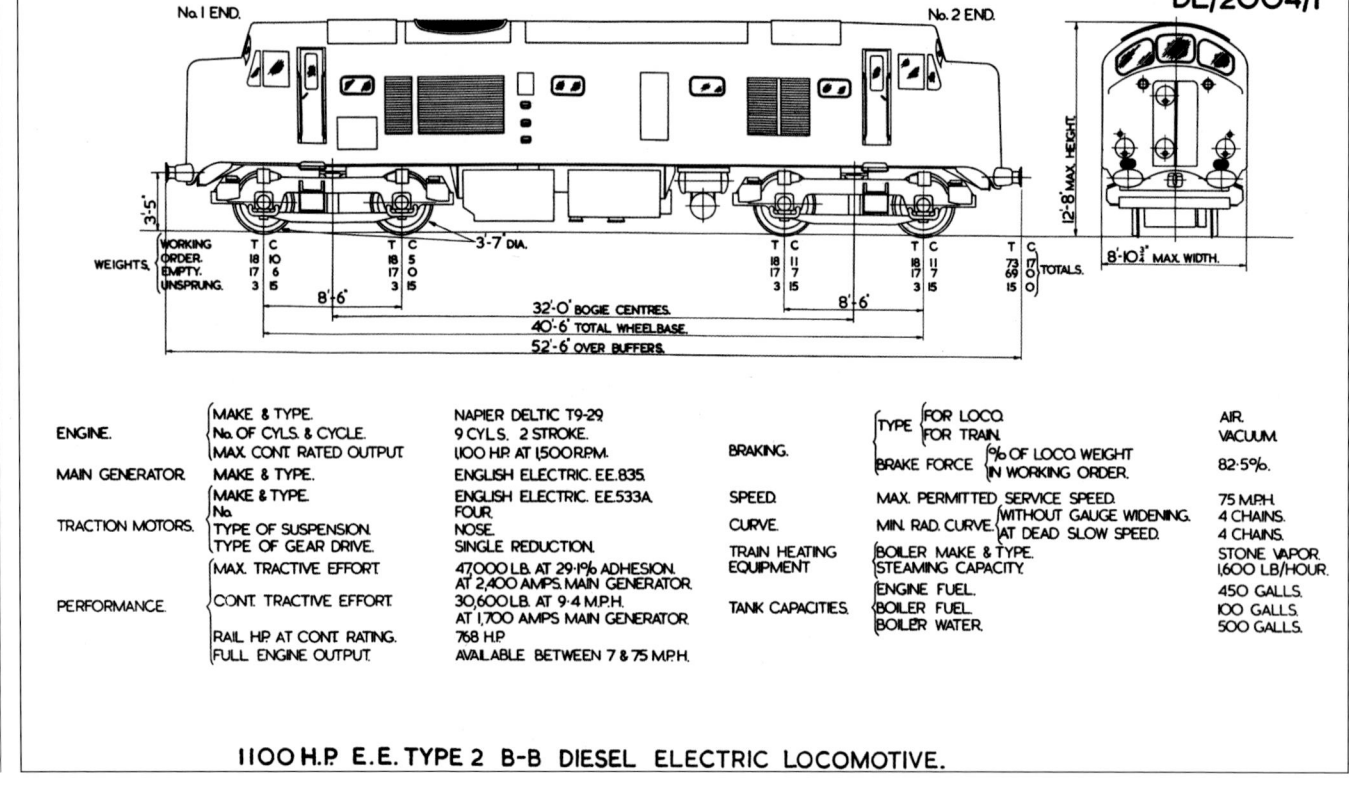

Diagram DE/2004/1. D5900-9 Side-Elevation. Shown fitted with oval buffers, four sandboxes per bogie (two per side), and no recesses or holes cut into bogie equalising beams, as originally envisaged. (BR Main-Line Diesel Locomotive Diagrams, 1961)

Technical Aspects • 21

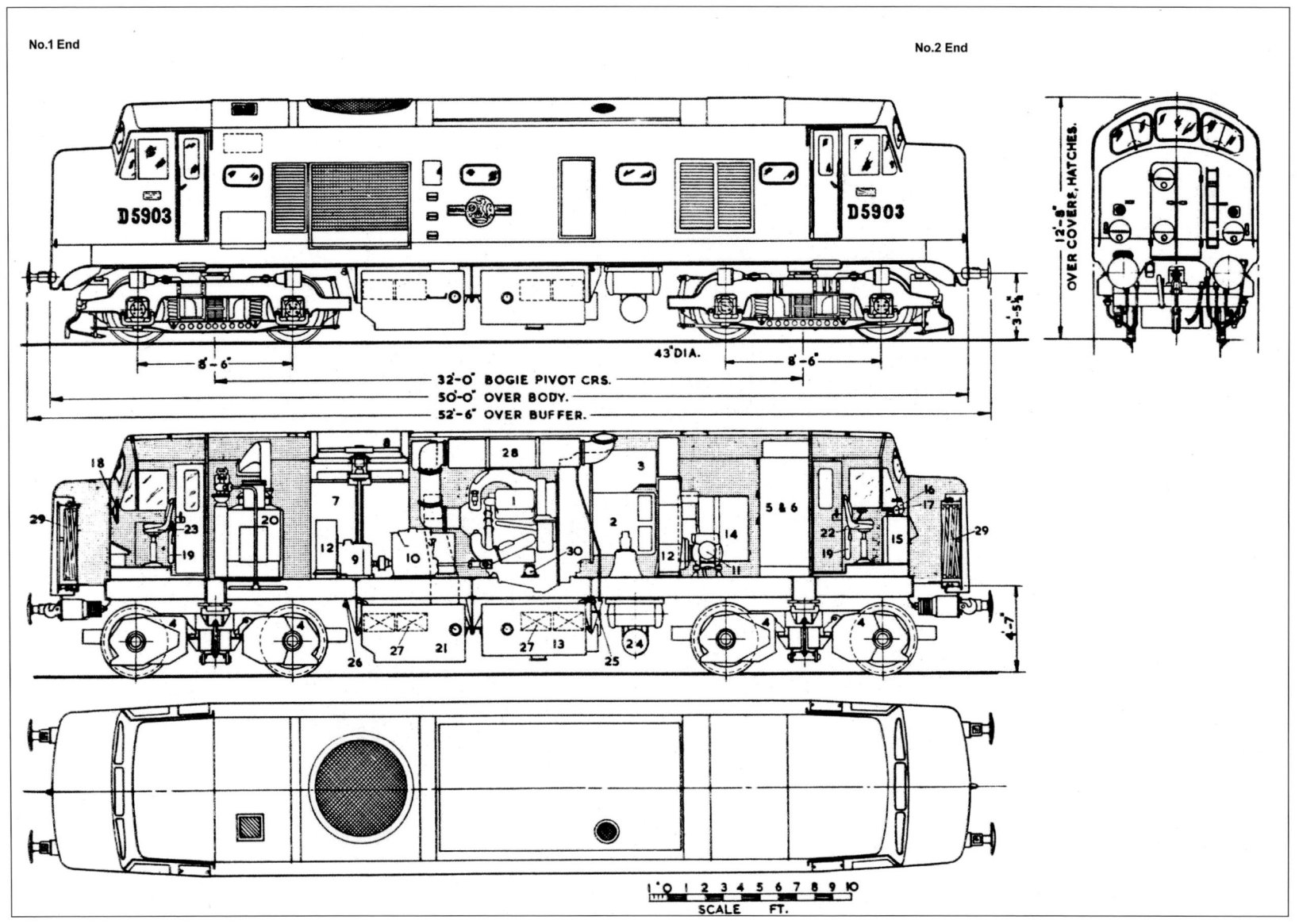

D5900-9 Side-, Top- and End-Elevations and Internal Layout Diagram. As delivered configuration: round buffers, two sandboxes per bogie (one each side) and holes cut into the bogie equalising beams. (BR Main Line Locomotive Layout Diagrams, Undated)

Equipment Layout: 1. Napier 'Deltic' engine; 2. Main generator; 3. Auxiliary generator; 4. Traction motors; 5. Control Cubicle; 6. Resistances; 7. Radiator Panels; 8. Radiator Fan; 9. Auxiliary Gearbox Drive; 10. Air Compressor; 11. Vacuum Exhausters; 12. Traction Motor Blowers; 13. Fuel Tank; 14. Air Filters; 15. Master Controller; 16. Vacuum Brake Valve; 17. Air Brake Valve; 18. Handbrake; 19. Fire Extinguishers; 20. Train Heating Boiler; 21. Water Tank; 22. Driver's Seat; 23. Assistant's Seat; 24. Air Reservoirs; 25. Fuel Tank Filler; 26. Water Tank Filler; 27. Batteries; 28. Silencer; 29. Gangway Connection; 30. Engine/Generator Flexible Mountings.

3.4 Diesel Locomotive Record Card

Diesel Locomotive Record Card for D5907. Note references to the 34/1100 and D11/1 classifications. (Courtesy: The National Archive, Kew)

Diesel Locomotive Record Card for D5907. Note reference to the officially declared periods in storage (1962/63 and 1968). (Courtesy: The National Archive, Kew)

3.5 Classification

1958

DML.34/1,100	DML: Diesel Main-Line 3: Deltic (Napier) engine 4: English Electric electrical equipment 1,100: Engine horsepower

W.e.f. 28/02/60

D11/1	D: Diesel 11: 1,100hp engine 1: Identifying number in "11" series.

W.e.f. 05/03/62

11/3	11: 1,100hp engine 3: English Electric

1968 (TOPS Reclassification)
Class 23

3.6 Underframe and Superstructure

The locomotive was a full width design and comprised a single central machinery compartment flanked by a cab at each end together with short nose compartments which incorporated gangway connections and some auxiliary equipment (including the compressor governor, CO_2 equipment, etc). The body shell adopted was the then typical EE 'bonnet' design with three windscreens, similar to the 'Pilot-Scheme' Type 4 fleet also being built by English Electric at VF at the time but suitably scaled down to suit the shorter overall length of the Type 2.

The underframe was of welded construction made up of two inner I-section beams and two outer channel sections, with intermediate welded cross-members and end drag boxes. The body structure was pre-fabricated, welded to the underframe and covered with welded medium-gauge steel plating. The cab 'bonnets' were also formed of steel plate.

The roof over the machinery compartment was removable in sections to facilitate installation and removal of the engine, generators, train-heating boiler, etc.. No cat-walk was provided on the roof.

External access to the cab was via side doors, with an extra body-side door on each side of the superstructure providing access to the engine room.

An internal walkway through the machinery compartment between the two cabs was provided, accessed via doors in the rear bulkhead in each cab. A slightly off-centre door from each cab gave access to the 'bonnet' section and the gangway doors.

Below left: D5909 Data Panel. (Anthony Sayer)

Below right: Vulcan Foundry, Newton-le-Willows. D5903 under construction. (Colin Marsden Collection)

A fuel tank, water tank and air reservoirs were hung from the underframe between the bogies. The fuel and water tanks were recessed on the '2-1' side of the locomotive to accommodate batteries. The fuel tank was provided with pressure filling connections, contents gauges and drain valves at either side of the locomotive.

Oleo pneumatic buffers were fitted.

3.7 Driving Cabs

The driving cab layout was of standard English Electric design with the driving position on the left and a non-driving 'secondman's' position on the right.

3.8 Napier 'Deltic' T9-29 Diesel Engine

3.8.1 Overview

Although originally designed to meet Admiralty marine requirements, the Napier 'Deltic' engine became an attractive choice for rail applications given their high power-to-weight ratio.

Sixteen T9-29 'Deltic' engines were built by D. Napier & Son, Netherton, Liverpool, with serial numbers 363, 370-80 and 385-8, although it is understood that 363 and 385 were dismantled fairly 'early on' for spares use. The sixteen comprised ten installed engines, three spares and three additional EE/Napier-owned spares, although another source suggests eleven BR engines (including one spare) and five EE/Napier spares.

3.8.2 Key Features

Deltic T9-29 engine. Model description: T = Turbocharged; 9 = No. of cylinders, 29 = Napier design reference. An alphabetic suffix was frequently used to denote the type of driven attachment; in theory, the 'Baby' Deltic engine should have been T9-29B (where B = flange mounted generator). In reality, 9-29 was sufficient as the Napier design reference in itself reflected a locomotive application with turbo-charging and coupled to a flange-mounted generator.

T9 'Deltic' engine, Stratford DRS, 19 March 1977. This and the following picture illustrate the engines removed from D5905 and D5909, being used as potential sources of spares for D5901 at Derby RCD. This view shows the free-end of the engine with the turbo-charger and associated pipework. (Tony Skinner)

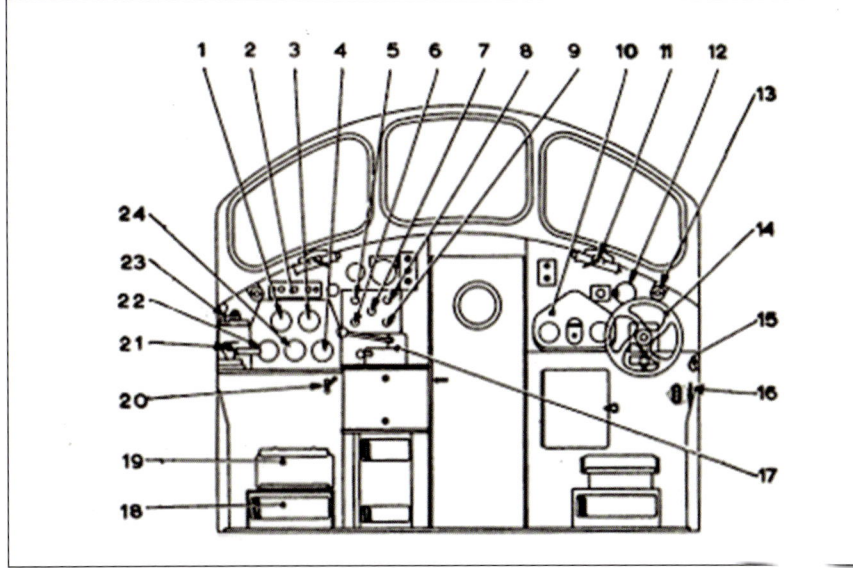

Driving Cab Equipment Layout. (English Electric Training Manual)
1. Speedometer; 2. Indicator lights; 3. Driving ammeter; 4. Duplex air pressure gauge (brakes); 5. Sanding button; 6. Exhausters speed button; 7. Regulator hold button; 8. Engine start button; 9. Engine stop button; 10. Boiler control panel; 11. Windscreen wiper; 12. Pull handle cover (fire fighting equipment); 13. Windscreen wiper control valve; 14. Hand brake handwheel; 15. Deadman's push button; 16. Vacuum release valve; 17. Controller; 18. Heater; 19. Deadman's pedal; 20. Horn valve; 21. Driver's air brake valve; 22. Main reservoir pressure gauge; 23. Driver's vacuum brake valve; 24. Duplex vacuum gauge.

T9 'Deltic' engine, Stratford DRS, 19 March 1977. Drive end of the engine, with the main generator and over-mounted auxiliary generator and No.2 bogie traction motor blower dwarfing the 'Deltic' engine itself. (Tony Skinner)

'Deltic' T9-29 Engine: Key features:

Description: Two-stroke, opposed-piston, compression-ignition, turbo-blown, liquid-cooled engine.

Cylinders: Nine, three banks of three cylinders formed in triangular configuration.

Overall dimensions:
- Length: 86.25in
- Width: 71.00in
- Height: 90.25in

Cylinder Bore: 5.125in
Piston Stroke: 7.25in (2 pistons per cylinder)

Piston speed at 1600 crankshaft rpm: 1932ft/min.

Swept volume: 2,692in^3 total.

Rated output: 1,100bhp at 1600 crankshaft rpm.
Engine speed continuously variable between 600 and 1600rpm.

BMEP at rated output: 101lb/in^2
BMEP = Brake Mean Effective Pressure (i.e. a performance metric used to compare different engines). BMEP measures the average (mean) cylinder pressure applied across the pistons to produce torque at the crankshaft through one complete cycle of operation.

Power per cylinder: 122.2hp.

3.8.3 'Deltic' Engine Characteristics

The 'Deltic' engine represented an integration of three identical two-stroke, opposed-piston, uniflow-scavenged engine banks to form a compact triangular unit. The triangular cross-section end view gave rise to the 'Deltic' name derived from the inversion of the Greek letter 'Delta' (Δ).

The engines used on BR consisted of either six banks of three cylinders (with two of these D18-25 units used in the production 'Deltic' locomotives (D9000-21)), or, three banks of three cylinders (one such T9-29 engine being deployed in the 'Baby Deltic' locomotives).

Each 18-cylinder engine in the production Type 5 'Deltics' was set at 1650hp, whereas the 9-cylinder unit in the 'Baby Deltics' was set at 1,100hp.

Locomotive engines were deliberately derated versions of the marine equivalents to facilitate longer intervals between overhauls; for the 1650hp-rated 18-cylinder engines such intervals were initially set at 4,000hrs with the rating calculated on the basis of thermal stresses within the engine. Similar marine engines were rated at 2,500hp with 1,000hrs overhaul intervals.

The 9-cylinder engines were similarly derated to a level substantially below their intrinsic capability to achieve an improved operating life, although the more conservative level of derating was consistent with the design requirement for a Type 2 1,100hp locomotive. The maximum hp per cylinder rating of the 'Baby Deltic' T9-29 engine at 122.2hp was, therefore, significantly greater than the D18-25 engine in D9000-21 (91.7hp), reflecting the different fuelling rate, the use of turbo-charging on the 'Baby Deltic' engines (as opposed to mechanically-driven scavenge blowers on D9000-21) and the higher crankshaft speed (1600rpm, as opposed to 1500rpm on the large 'Deltics').

The compact dimensions and low weight of the engine were functions of the inherent design characteristics of the engine rather than an excessively high cylinder output level. The compactness of the 'Deltic' engine in relation to its power was most noticeable from the limited amount of locomotive body space which it occupied, particularly when compared with the bulky main generator or the train heating boiler.

Another method of comparison was the engine weight in lb per single horsepower produced; thus the D18 engine was 6.1lb/hp, the turbo-charged T9 6.8lb/hp, both substantially superior to the 16SVT engine used in English Electric's Type 4 'Pilot-Scheme' design (D200-9) of 20.4lb/hp.

3.8.4 'Deltic' Engine' Design

The T9 'Deltic' engine comprised four main assemblies: the main triangle (formed of cylinder blocks, crankcases, etc.) (see Sections 3.8.5-11), the phasing gears at the driving end of the assembly (3.8.12 and 13), the turbo-charger unit (3.8.15) and the integral gear box at the free end (3.11.1).

3.8.5 Cylinder Blocks

Three identical light-alloy cylinder blocks, each containing six pistons, were arranged so that in end view the configuration of the engine was an inverted triangle. Each bank of cylinders consisted of three cylinder blocks giving nine cylinders overall (containing a total of eighteen pistons). A crankcase was situated at each apex of the triangle giving three crankcases in total (see diagram on page opposite).

The triangular assembly of cylinder blocks and crankcases formed the basic structure of the engine held together by long high-tensile bolts which passed through drillings in the cylinder blocks. These through-bolts carried all combustion loads, the cylinder blocks remaining in compression. The bolts collectively united the whole assembly into a rigid structure.

The cylinder blocks incorporated the passages for turbo-charged air and exhaust gases, plus the coolant galleries. Air inlet ducts, connected to the turbo-charger, were cast along the inward and outward facing sides of each cylinder block, and an exhaust manifold was attached to a mounting face parallel with the air duct on the outward side.

Between the air inlet ducts and the exhaust manifold on each block, a mounting face was machined for the attachment of a camshaft casing, and between this face and the exhaust manifold, apertures were provided for the fuel injectors (one per cylinder).

The triangular arrangement facilitated the opposed-piston design enabling each crankshaft to serve two adjacent cylinders, two connecting rods being attached to each crank throw. As a consequence there were no cylinder heads on a 'Deltic' engine, compression being achieved by the opposing pistons.

The engine was mounted on two feet bolted to two extensions cast integrally with the bottom crankcase on either side.

3.8.6 Cylinder Liners

The six wet liners in each cylinder block were machined from hollow steel forgings and were chromium plated in the bore to reduce piston wear, the major portion of the working surface being covered by etched 'dimples' to retain the lubricant. Externally, each liner was machined to provide four recesses which mated with similar recesses in the cylinder block to form annular cooling jackets.

A flange at the inlet end of the liner seated onto a shoulder in the cylinder block to provide longitudinal location, the assembly being secured by a ringnut.

The air ducts and exhaust manifold on the cylinder blocks married-up with the respective ports on the cylinder liners by way of chambers formed round the liners in the casting. In the 'DELTIC' engine, nine exhaust ports were arranged around the exhaust end of the liner, and fourteen inlet ports around the inlet end; it is presumed that a similar multiple-port arrangement applied with the T9 engine.

3.8.7 Crankcases

The crankcases, as viewed from the free end of the engine were: 'AB' top-left, 'BC' top-right, 'CA' bottom. Therefore, cylinder bank 'A' was to the left, 'B' was across the top of the triangle and 'C' was to the right.

The three crankcases were one-piece light-alloy castings. Each casting was assembled between adjacent cylinder blocks by two faces machined at 120°. The outward facing side of each crankcase was closed with a light-alloy cover that gave access to the main and big-end bearings.

The two top crankcases were broadly similar but the lower crankcase was deeper in section to provide a sump for drain oil for the 'dry sump' lubrication system with an outlet for scavenge oil incorporated in the wall on the 'C' side. The free end of the 'CA' crankcase was closed by a casting that housed the fuel filters and provided a mounting face for the coolant and fuel pumps.

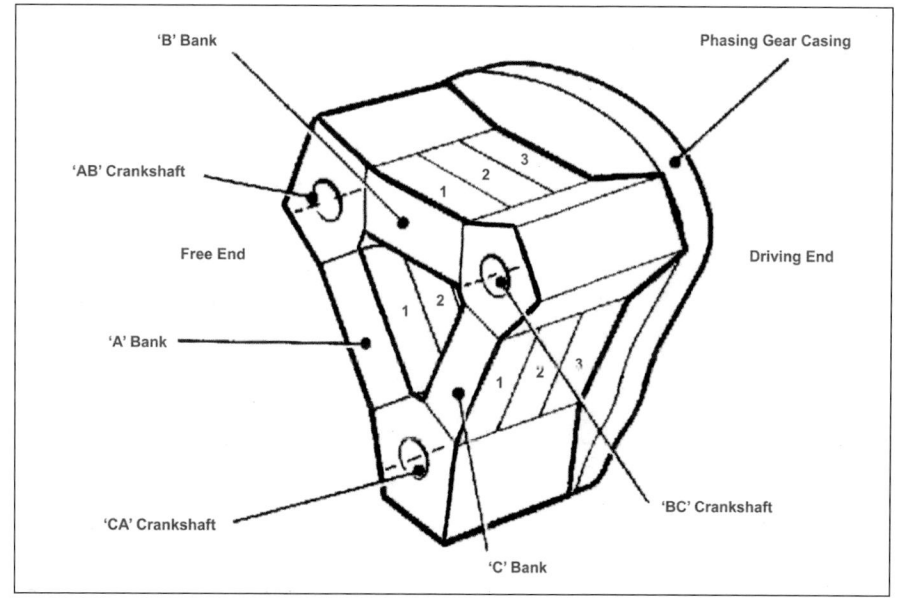

T9 'Deltic' Engine Cylinder Block and Crankshaft Arrangement.

Internally, each crankcase was strengthened by stiffening bulkheads between which were located the crankshaft bearing housings. The bulkheads were reinforced with high-tensile tie-bolts, one bolt passing through each bulkhead and crankshaft bearing cap. Passages in the crankcase walls and bulkheads conveyed oil to the main bearings and to the drive gears and their bearings.

Four thin-wall bearings supported each crankshaft. Each bearing comprised two steel halves lined with lead-bronze, lead-coated and indium infused. One half bearing was located in a housing machined in each crankcase web and was held in place by a bearing cap which contained the other half-bearing and was drilled to accommodate the crankcase tie-bolts.

3.8.8 Crankshafts

Each crankshaft was machined from an alloy steel forging nitrided all over, the bearing surfaces being finished by lapping. Suitable drillings in the crankwebs, journals and crankpins conveyed lubricating oil from the main bearings to the big-end bearings.

The two top crankshafts ('AB' and 'BC') were identical and rotated in a clockwise direction when viewed from the free end. The bottom crankshaft ('CA') rotated in an anti-clockwise direction so that the correct phase relationship was achieved. Each of the three crankshafts drove their respective phasing gears in the phasing gear casing (see Section 3.8.12).

The power transmitted through each crankshaft was identical, with any torsional vibration controlled by the use of quill shafts at the drive end in tune with Holset viscous dampers at the free end.

3.8.9 Pistons

The Mk.3 pistons on a 'Deltic' engine were assembled from two separate components i.e. an outer aluminium piston body with an unbroken cylindrical skirt combined with a copper-alloy 'Hidural' crown, and an inner gudgeon-pin housing. The two components were held together by an internal taper-seated circlip.

The 'Hidural' crown was necessary because of the high exhaust temperatures generated in two-stroke engines; in four-stroke engines only alternate strokes are firing strokes giving the pistons an opportunity to cool. The superior thermal qualities of copper were more suited to the two-stroke arrangement, hence the use of a separate crown.

Three compression rings were fitted in grooves immediately below the piston crown, the two uppermost grooves being machined in an austenitic iron insert bonded to the aluminium body by the Alfin process to reduce ring groove wear. Two scraper rings at the bottom of the piston skirt provided the requisite oil control by removing excess oil from the cylinder walls. The excess oil was transferred to the inside of the piston through the grooved and slotted lower scraper ring and through holes drilled from the inside of the ring groove to the inside of the piston.

The gudgeon-pin housing carried a fully-floating gudgeon-pin in bronze bearing bushes. When the piston components were assembled, external grooves machined in the top and sides of the gudgeon-pin housing were covered by the piston body to form ducts in which oil, supplied from the small end of the connecting rod, was shaken by the reciprocation of the piston, thereby transferring heat from the piston crown to the skirt.

3.8.10 Connecting Rods

Each crankpin carried a pair of connecting rods, one forked and one plain, manufactured from drop-forgings machined and polished all over.

Each exhaust piston was attached to a forked rod with a steel shell at the large end. The inner surface of the shell had a thin-wall bearing for the crankpin.

Each inlet piston was attached to a plain connecting rod, the large end having a thin-wall bearing. The plain rod oscillated inside the forked connecting rod. Each forked big-end provided the journal for the plain rod big-end-bearing on the outside.

The rods were drilled lengthwise to transfer oil from the big-end bearings to the small-end bearings for gudgeon-pin lubrication and piston cooling. Since the bottom crankshaft rotated in the opposite direction to the top crankshafts, a different arrangement of oil grooves and holes was provided in the big-end bearings.

3.8.11 Camshaft Casings and Camshafts

Attached to the side of each cylinder block was a camshaft casing containing the injection pump camshaft and its drive, mountings for the injection pumps, and integral galleries for the circulation of fuel to the pumps. Each camshaft was supported in plain bearings fed with lubricating oil from the inside of the hollow camshaft.

Two bevel gears in the camshaft casing, one keyed to the driving end of the camshaft, were driven from the adjacent crankshaft gear by a train of spur and bevel gears in the Phasing Gear Casing; the gears in the camshaft casing and those in the phasing gear housing being connected by a short quill-shaft enclosed in a split tubular casing.

Coupling serrations at each of the quill-shafts were arranged in a vernier combination to enable adjustments to be made when timing fuel injection. A graduated timing segment, which could be viewed through an inspection port, was fitted to the free end of each camshaft.

3.8.12 Phasing Gears

Attached to the driving end of the Napier engine assembly was the Phasing Gear Casing, in which were housed the train of gears which transmitted the drive from the three crankshafts to various other components.

The phasing gear casing consisted of a pair of similar light alloy castings, bolted together in a vertical plane. The free-end casting had external machined faces for attachment to the engine and the driving end casting had an external machined face for the main generator adaptor ring. Both castings incorporated bosses for the accommodation of housings for the gear trains and output shaft bearings, together with passages for the circulation of oil.

Throughout the phasing gear casing there were a number of 'sparge jets'. These were critical in the lubrication of the gears due to there being no sump for gear oil as in a conventional gearbox. Each jet assembly had two or three holes drilled into a machined face and was arranged to spray oil directly onto the gears.

As already mentioned, the three crankshafts, incorporating quillshafts and spur gears at their drive ends, transferred the drive from each crankshaft into their respective gears within the Phasing Gear Casing. Phasing adjustments of approximately ¼°. could be obtained by means of a vernier arrangement of two gear-toothed couplings one at each end of each quill shaft.

Phasing gear trains, including idler gears, brought together the output from the three crankshafts into one common output shaft on the centre line of the engine which drove the main generator. The gears in the 'Baby Deltic' phasing case were arranged so that the main generator turned at engine speed, unlike the D18-25B in the production 'Deltics' where the generator speed was three-quarters of the engine speed. The output to the main

'Deltic' T9-29 Engine Gear Trains. (Derived from R.M. Tufnall [1979])

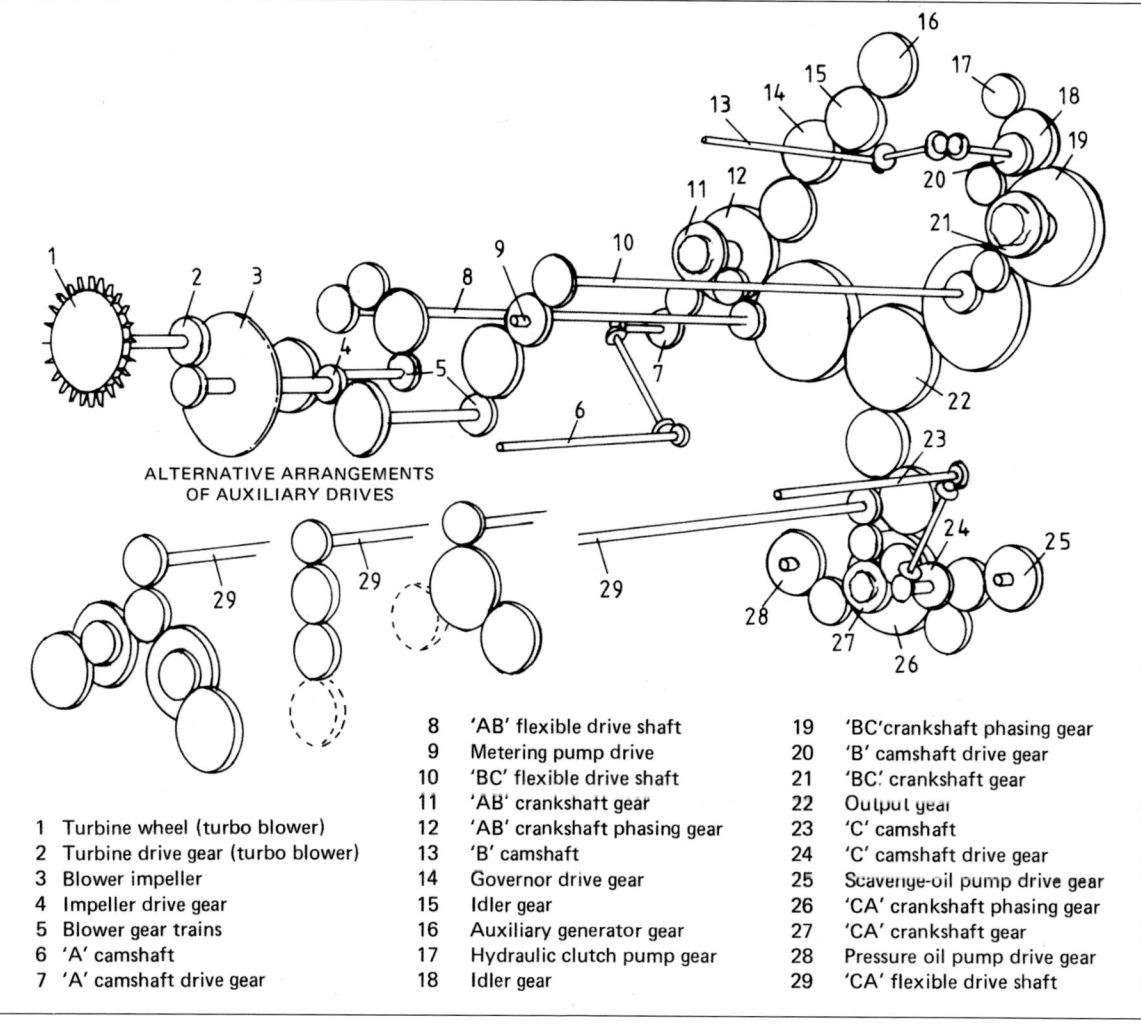

ALTERNATIVE ARRANGEMENTS OF AUXILIARY DRIVES

1	Turbine wheel (turbo blower)	8	'AB' flexible drive shaft	19	'BC' crankshaft phasing gear
2	Turbine drive gear (turbo blower)	9	Metering pump drive	20	'B' camshaft drive gear
3	Blower impeller	10	'BC' flexible drive shaft	21	'BC' crankshaft gear
4	Impeller drive gear	11	'AB' crankshaft gear	22	Output gear
5	Blower gear trains	12	'AB' crankshaft phasing gear	23	'C' camshaft
6	'A' camshaft	13	'B' camshaft	24	'C' camshaft drive gear
7	'A' camshaft drive gear	14	Governor drive gear	25	Scavenge-oil pump drive gear
		15	Idler gear	26	'CA' crankshaft phasing gear
		16	Auxiliary generator gear	27	'CA' crankshaft gear
		17	Hydraulic clutch pump gear	28	Pressure oil pump drive gear
		18	Idler gear	29	'CA' flexible drive shaft

generator was driven off a flange-mounted shaft attached to the centre gear in the phasing gear casing.

A separate gear train provided the drive for the auxiliary generator (via the 'AB' crankshaft phasing gear).

Secondary gear trains provided the drive for:

- the three fuel injection pump camshafts (driven from their three associated crankshaft phasing gears ['AB', 'BC' and 'CA']),
- the Ardleigh governor (driven from the 'AB' crankshaft phasing gear through an idler gear and a driven gear and quill shaft),
- the turbo-charger (driven from the top two crankshaft phasing gears ['AB' and 'BC'] via flexible shafts),
- the scavenge and pressure-oil circulating pumps (via the 'CA' crankshaft phasing gear),
- the free-end gearbox (driven from the bottom crankshaft phasing gear ['CA'] via a flexible shaft; the free-end gearbox, in turn, drove the auxiliary gearbox [see Section 3.11]).

The flexible drive shafts which ran the whole length of the engine from the driving-end to the free-end were supported in plain bearings in each of the angles between the adjacent cylinder block mounting faces.

The complex arrangements of gear trains are illustrated on page 29.

3.8.13 Phasing and Timing

The correct crankshaft phasing relationship and timing of the 'Deltic' opposed-piston engine was essential, demanding precise initial set-up and subsequent maintenance. The 'phasing process' was a lengthy procedure in which all the crankshafts' angular positions were set relative to each other (and, therefore, the positions of the inlet and exhaust pistons) before each camshaft was then set relative to its associated driving crankshaft (for fuel pump operation).

In two-stroke engines, with piston-controlled inlet and exhaust ports, the exhaust port has to be opened before the inlet port to ensure maximum efficiency in scavenging and re-charging. This exhaust lead was normally obtained by the positioning of the ports in relation to the piston movement, but in opposed-piston engines it could be obtained by a combination of port positioning and crankshaft phasing, the crankshafts being phased so that the inlet and exhaust pistons did not reach 'top dead centre' together. The difference between the relative positions of the two pistons was termed the 'phase difference'. In the 'Deltic' engine, the phase difference was 20° of crankshaft rotation, and the phasing gearing was so arranged that this angle was applied to all cylinders and the relative movement of the pistons in all cylinders was identical.

As the exhaust and inlet pistons in each cylinder were out of phase, port timing could not conveniently be stated in terms of degrees of rotation of the crankshaft to which each piston was connected; instead,

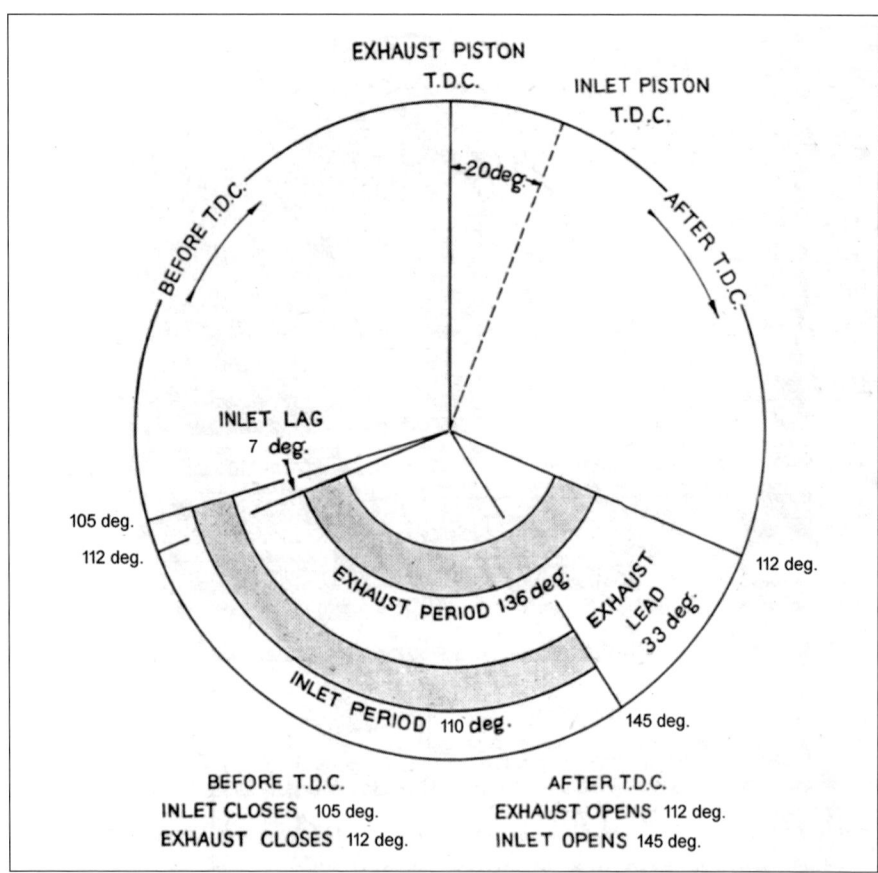

Napier 'Deltic' T9-29 Timing Diagram.
T.D.C. = Top Dead Centre. (Derived from 'The Napier Deltic Engine', *Diesel Railway Traction*, December 1955, with angles modified to reflect the T9-29 engine)

the exhaust piston was taken as the datum, and the timing of both pistons referred to the crankshaft to which the exhaust piston was connected. The ideal exhaust port lead for this type of engine was that which allowed the combustion pressure to fall to the scavenge pressure at the point of inlet port opening. On the 'Deltic' T9 engine this lead was determined to be 33°, and therefore 13° lead had to be obtained by port positioning in addition to the 20° phase difference between the inlet and exhaust pistons imposed by the engine geometry. The inlet port was closed 7° after the exhaust port, and thus the turbo-blower supercharged each cylinder after the exhaust port had closed. During the period that both ports were open, the air flow from the turbo-charger scavenged the cylinder of all exhaust gases and cooled the inside of the cylinder (the so-called 'uni-scavenge' system).

With the three crankshafts set relative to each other, the camshafts had to be set to enable the accurate timing of fuel injection, and was, therefore, equally critical as crankshaft phasing.

3.8.14 Fuel Injection System

Fuel was taken from the locomotive fuel tank by a Varley DH25 fuel transfer pump, which was powered by an EE type TFZ97918 fractional horsepower motor. The transfer pump circulated fuel at low pressure through the diesel engine fuel rails, and it was from these that the fuel injection pumps drew their suction.

The EE motor was initially supplied with current from the batteries when starting, but thereafter it ran continuously on current supplied by the auxiliary generator.

The pumps were of CAV manufacture modified by Napier to suit the design requirements of the engine, and were mounted on the camshaft casings extending the length of the cylinder blocks; the pumps were operated by camshafts indirectly driven from their respective crankshaft phasing gears.

Each fuel injection pump supplied fuel at extremely high pressure to one fuel injector, one for each cylinder. Each fuel injector assembly comprised a nozzle and a nozzle holder. At the designated 'break open' pressure, the fuel pump delivered fuel to the injector nozzle and injected it into the cylinder at the prescribed moment for combustion to begin; it should be noted that the peak injection pressure significantly exceeded the 'break open' pressure of the valve in the fuel injector nozzle. The injectors delivered fuel at the outer edge of the combustion chamber, the fuel being transmitted to the centre by the swirl of air from the inlet ports.

The fuel circulating pump had an output in excess of the engine's fuel requirements, surplus fuel being returned to the tank through a pressurising valve. This ensured that a constant supply was maintained at all injection pumps and also prevented the formation of air pockets in the system.

3.8.15 Turbo-charger

Unlike the scavenge-blown engines in the production 'Deltics', the T9 engines in the 'Baby Deltics' had an exhaust-gas driven turbo-charger group mounted on the free-end of the engine. The turbocharger in the Baby Deltics was unique in that it was incorporated into the engine gear train rather than being a free entity as in most locomotive applications.

The installed Napier turbo-charger was initially driven by two flexible shafts connected by gears to the 'AB' and 'BC' crankshaft phasing gears via a speed-raising gearbox. The turbo-charger was then augmented at higher engine output levels when the exhaust gases progressively acted on the turbo-charger turbine blades in the conventional way, relieving the torque load on the drive shafts in the process.

This arrangement was all part of addressing the massive problem with all port-controlled two-stroke engines, that of getting sufficient air into the engines, initially for starting, but also during the normal working cycle when running.

Three outlets from the turbo-charger fed the air-inlet ducts in the cylinder blocks.

3.8.16 Exhaust

The exhaust from the three cylinder banks combined in a lagged saddle tank and subsequently piped to the silencer.

The exhaust manifolds were cooled by a jacket through which engine coolant was circulated.

3.8.17 Oil Lubrication

Oil lubrication was on the dry sump principle. Being of triangular form the 'Deltic' engines were somewhat top heavy. To keep the engine height to a minimum they did not have a conventional sump into which the oil drained and was stored; instead, a separate tank was employed, located adjacent to the engine.

The use of an external tank required two pumps. Firstly, there was a main-engine driven pump which took oil from the tank and pressurised it prior to delivery to the 'main pressure service' (via a pressure filter), which supplied oil to the three crankcases to lubricate the principal bearings and the pistons; from the 'main service', oil was fed at reduced pressure to the phasing gears and to the auxiliary gear trains in the crankcase. Oil was also supplied from the 'main service', through a restrictor, to the scavenge blower gear train and to the blower impeller shaft bearings.

From the various galleries within the engine, gravity returned the oil to the 'CA' crankcase, from where a scavenge pump drew it via a filter back to the oil tank. This system was referred to as the 'pressure/scavenge system'. Both pumps were located on the engine side of the phasing case, one either side of the 'CA' crankcase. On top of the oil scavenge pump was a hose which carried the oil from the pump to the cooler from where it was returned to the tank. The oil inlet temperature to the engine was thermostatically controlled.

The 'Deltic' engines had no motorised lubricating oil priming pump for use prior to starting and this function had to be undertaken manually by a hand pump. When starting the engine from cold, a three-way thermostatic valve isolated the radiators thereby assisting the engine to rapidly attain normal working temperature.

Valves were provided for draining the radiators and oil tank, and a drain cock was fitted to the engine sump.

3.8.18 Cooling System

Power unit oil and water in closed circuits were air-cooled by a roof-mounted fan which drew air through double-bank Marston Excelsior radiators. Manually operated Spiral Tube shutters were fitted over the radiator panels on the bodyside. Temperature was thermostatically controlled.

The radiator fan was mechanically-driven from the auxiliary gearbox, which in turn was driven from the 'CA' crankshaft via gears and flexible shafts. The fan itself had eight blades (subsequently increased to twelve).

When starting the engine from cold, a three-way thermostatic valve initially isolated the radiators from the system to assist the engine to reach its normal operating temperature.

Suitably softened and corrosion-inhibited cooling water was circulated around the cylinder blocks, exhaust manifolds, and radiators.

A header tank was incorporated in the system to make up for water losses. A water gauge was fitted to the tank and the system was normally filled through cocks located one at either side of the locomotive directly beneath the radiators. A hand pump and suction hose was also provided, to facilitate the addition of chromate solution and to enable the system to be filled in an emergency when no pressure supply was available.

External air for engine room ventilation and for the traction motor blowers passed through bodyside grilles with panel filters of the oil-wetted type and expelled through the roof.

An internal EE Memo from the Rolling Stock Design Department, Preston, to Napier, Acton, dated 25 June 1956, highlighted some of the difficulties associated with the design of the radiator area:

D5909 Engine Room. Napier 'Deltic' engine removed and looking towards No.1 end from the vacant space. The large pipe descending from the roof is the (disconnected) engine exhaust outlet pipe. The large roof-mounted 'lump' at the top of the picture was the silencer. The rectangular tank in the lower middle part of the picture is the engine lubricating oil tank. The mechanically-driven compressor, to the left of the oil tank, is just visible. The double-bank Marston Excelsior radiator is clearly visible in the middle distance; the radiator fan drive is just visible within the radiator 'tunnel'. (Anthony Sayer)

BTC 1,100hp DE Locomotives – Arrangement of T9 Engine
Due to restrictions imposed by the BTC the locomotive structure is narrower than that used on the… experimental 'DELTIC' locomotive. Also due to the necessity for accommodating sufficient fuel below a very much shorter locomotive we have been forced to mount the power unit at an increased height from rail. The radiators will, therefore, be placed in the superstructure sides as per our more usual practice.

The oil filters… are now a serious obstruction to the gangway on each side of the locomotive leaving a clear passage only 10¼" wide at thigh level. Would you therefore please consider the location of the filters and arrange so that no portion of the engine projects beyond the crank chamber covers in width. If the filters can be mounted vertically, or even at a smaller inclination, the gangway would be much improved.

Quite how this issue was resolved is unknown, but the photograph opposite illustrates the limited space available for through passage in the radiator area .

3.9 Principle of Operation of 'Deltic' Engines

The excellent and highly informative books produced by the 'Baby Deltic' Project, *'Baby Deltic': The Story of an Engine's Rebirth* (2009 and 2023), describe the key principles of operation of the T9-29 engine, as follows:

- Each cylinder contained two pistons connected at the big-end to a crankshaft.
- Each piston controlled the uncovering and covering of its relevant inlet or exhaust port.
- The exhaust piston 'led' the inlet piston by 20° of crankshaft rotation; the exhaust piston, therefore, covered the exhaust port before the inlet piston covered the inlet port enabling a small degree of pressure-charging of the cylinder to occur.
- As the pistons came together, and after the inlet port was covered, the trapped air in the cylinder was compressed and its temperature rapidly increased.
- At a point about 30° before the exhaust piston reaches top dead centre (i.e. the limit of its stroke) fuel injection commenced.
- The fuel injected into the cylinder ignited immediately and created a massive pressure rise which forced the pistons apart.
- Combustion was completed by the time the exhaust piston uncovered the exhaust port which allowed spent gases to pass to atmosphere.
- Once the inlet piston had uncovered the inlet port, air was free to pass through the cylinder from the inlet to the exhaust port, enabling both cooling and scavenging of the cylinder prior to the whole cycle repeating.

3.10 Advantages and Disadvantages of the 'Deltic' Engine over 4-Stroke Engines

The publicity of the day waxed lyrical about the advantages of the two-stroke opposed-piston 'Deltic' engines. Much of this material was routed via various railway magazines and professional journals and was ultimately sourced from the manufacturers themselves i.e. English Electric and Napier. It is not surprising, therefore, that the 'Deltic' engine was described in very glowing terms. For example, C.D. Carmichael in his paper 'The Design and Development of the Deltic Engine' (Manchester Association of Engineers, November 1956), '… the opposed-piston arrangement as used in the [prototype] 'DELTIC' possesses such great technical merits when designed in triangular form that it displays overwhelming advantages for the two-stroke system.'

Given below are some of the advantages offered by the 'Deltic' engine as put forward by EE and Napier:

- High power/weight ratio. By its very nature a two-stroke engine (with one power stroke in two) is potentially capable of producing more power than a four-stroke unit of equivalent size and weight (one power stroke in four). Given that the 'Baby Deltic' design objective was to produce an engine of Type 2 capability within a 72ton overall weight limitation, this factor represented a key point in favour of the two-stroke engine.
- Engine simplicity – avoidance of valve gear. A high power/weight ratio required an engine of relatively small cylinder dimensions running at high crankshaft speeds. Four-stroke engines required valve gear of some kind, whereas with two-stroke engines this was avoided by using piston-controlled ports.

In the 'Baby Deltic' the two-stroke engine, in opposed-piston form, was considered to offer a higher degree of simplicity of operation, although it was recognised that this advantage was at least "partially offset" by the requirement to achieve and maintain accurate phasing of the three crankshafts and associated gear trains to provide an optimised output.

- Thermal efficiency. The cylinder combustion chamber in the T9-29 opposed-piston engine, formed by the space between the two pistons at their innermost positions, was thermally very efficient given that the amount of cool combustion chamber wall was minimised thereby reducing liner heat losses.
- Effective piston stroke. The opposed-piston phasing arrangement enabled practically the complete circumference of the cylinder wall to be made available for inlet and outlet ports, incorporated without undue sacrifice of effective piston stroke.
- Gas flow efficiency. The fact that the scavenge air entered the cylinder at one end and left at the other, the 'uniflow' scavenge system, enabled burnt gases to be efficiently expelled by the incoming air without undue mixing of the two, whilst also providing a degree of pressure-charging. It was claimed that the efficiency of the scavenging and charging processes gave the opposed-piston engine significant advantages over other types of high-speed engines.
- Pressure/turbo-charging. The inlet ports remained open for a short while after the exhaust ports were closed thereby allowing the maximum quantity of air to be packed into the cylinder.
- Scavenge air efficiency. The air inlet ports around the whole circumference of the cylinder liners with a partly tangential direction of entry imparted a swirl to the incoming air enabling the distribution of injected fuel as uniformly as possible within the combustion chamber. It was claimed that this assisted scavenging and promoted efficient combustion, the engine timing enabling approximately 30 per cent of the charge to scavenge and cool the combustion space and piston crowns before compression commenced.
- Engine life and reliability. In most cases the life and reliability of a high-speed engine is poor compared with a low-speed engine, with higher rotational speeds resulting in increased wear and tear. However, it was claimed that the 'Deltic' engine possessed a number of advantages not available to the low-speed designs, mainly resulting from the reduced size of components used and the consequent possibilities for using higher-grade materials and improved techniques in manufacture. This involved high precision engineering (which Napier was renowned for) albeit at significantly greater cost. The extent to which this advantage manifested itself is, however, very difficult to quantify in pure cost/benefit terms.
- The small size and weight of the 'Deltic' engine facilitated component exchange at depots, a philosophy used extensively by the Western Region (i.e. direct replacement of complete defective components by new or overhauled ones to minimise locomotive downtime at Main Works). 'Repair by replacement' of 'Deltic' engines at Stratford Works was deployed extensively (more than anticipated as it turned out), but appropriate cranage of sufficient capacity at Finsbury Park maintenance depot was never installed despite strong representations by the local management to do so.

The reality of the situation was that in the late 1950s English Electric did not have a four-stroke design which could produce 1,100/1,250hp whilst conforming to the maximum 72t locomotive weight stipulation. So EE had to resort to the Napier 'Deltic' marine engine to resolve this issue and as a consequence had to make every effort to promote its two-stroke alternative.

The 'Deltic' design was promoted as being very simple operationally; conceptually this was true but it significantly underestimated the sheer technical complexity of the engine, the specialist components required in its construction (with all of the additional associated costs) and the difficulty of setting up and maintaining the phasing of the crankshafts and complex gear train arrangements. The criticality of the phasing appears to have been somewhat understated.

Napier produced some highly sophisticated and technically competent equipment for the Royal Navy, but the suitability of such complex engines in a railway environment has to be questioned,

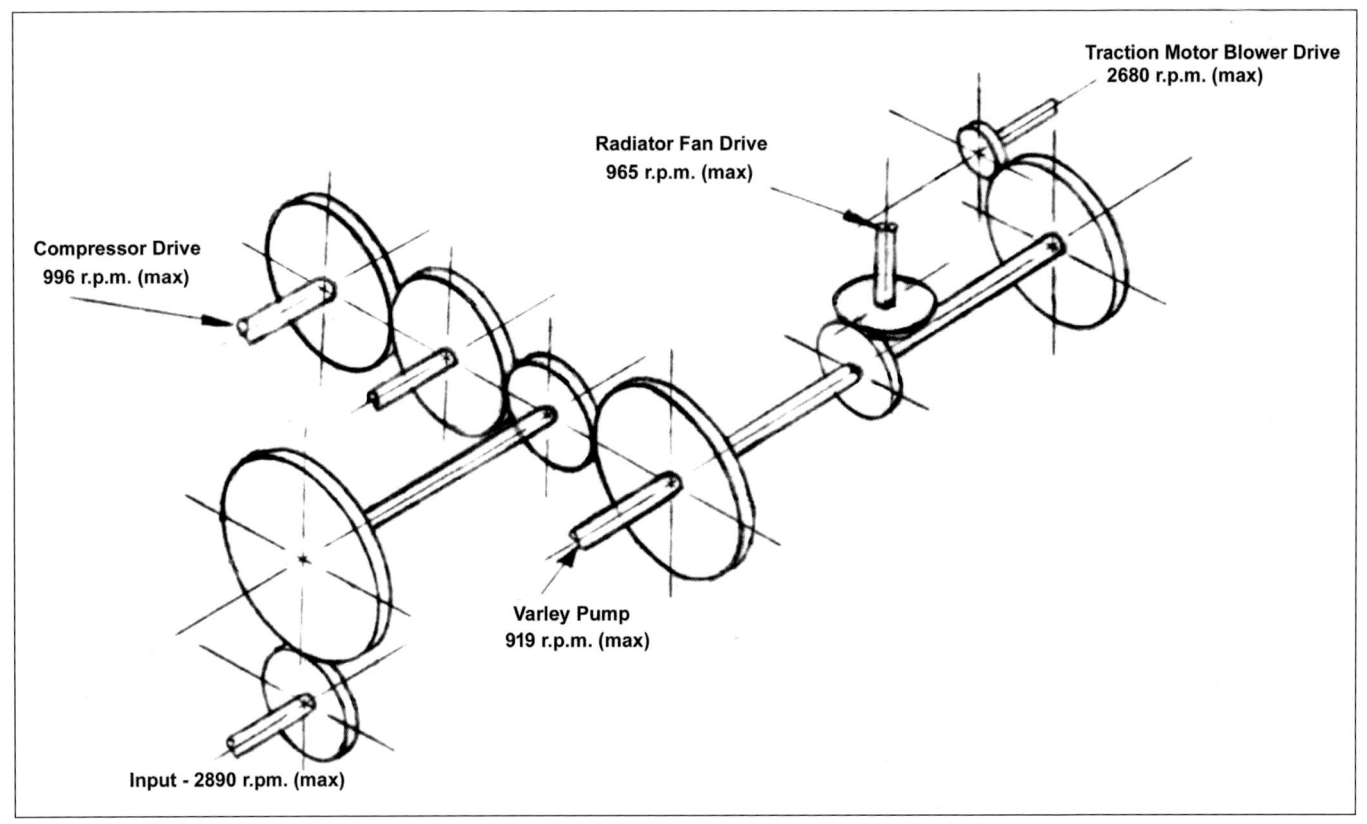

'Baby Deltic' Auxiliary Gearbox Arrangement. The power and rotational speeds of the various drive shafts were: Input: 60/68hp at 2890rpm; Radiator Fan: 30hp at 965rpm; Traction Motor Blower: 15hp at 2680rpm; Air Compressor: 23hp cold/15hp warm at 996rpm. (Derived from internal EE correspondence)

both in terms of the cut-and-thrust of day-to-day stop/start operations and against the background of steam-age maintenance standards and expertise.

It is important to note that most of the advantages offered by EE were not quantified. Not least of these was the relative power advantages of two-stroke versus four stroke. The two-strokes (one power stroke in two) were not twice as efficient as four-strokes (one power stroke in four) given the incomplete exhausting of burnt gases before new incoming air was admitted into the combustion space. Similarly, oil usage must have been high with the passage of burnt and unburnt lubricating oil into the exhaust, leading to the characteristic 'blue' fumes.

Issues experienced with the 'Deltic' engines are discussed in Section 9, both in terms of specific issues (most notably cylinder liner fractures and piston crown failures) and in general locomotive availability terms. However, the less tangible and difficult to quantify issues should not be forgotten e.g. maintenance challenges, cost of consumables, etc.

3.11 Mechanically-Driven Auxiliary Equipment

3.11.1 Free-end Auxiliary Gearbox

A gearbox located at the free-end of the engine was gear-driven by a shaft off the 'CA' crankshaft which passed along the length of the engine along the inside angle of the 'CA' crankcase exactly like the drive for the turbo-charger. The purpose of the free-end gearbox was to drive an auxiliary gearbox from which horizontal drives were taken to the Westinghouse air compressor (via a plate clutch), the Varley coolant circulating pump and the No.1 end traction motor blower; in addition, a vertical shaft with flexible couplings was taken to the radiator cooling fan via a centrifugal clutch.

3.11.2 Traction Motor Blowers

The traction motors were force-ventilated by two mechanically-driven blowers. No.1 traction motor blower, for cooling the two traction motors on the No.1-end bogie, was driven off the auxiliary gearbox and the No.2 blower, for cooling the traction motors on No.2-end bogie, also mechanically driven, this time from an extension of the auxiliary generator armature shaft. Both two traction motor blowers were Aerex Type 2TP040.

Each blower drew in air through filters and delivered it through ducting to the associated pair of motors on the adjacent bogie. Since both blowers were driven, indirectly, by the diesel engine, the quantity of air delivered to the traction motors was proportional to engine speed and, therefore, locomotive power output.

3.11.3 Air Compressor
As delivered, the Westinghouse air compressor (type 2EC72A) was horizontally driven off the auxiliary gearbox via a plate clutch. It should be noted that after refurbishment the compressor was substituted by an electrically driven Westinghouse (type 2EC38B) in order to relieve load on the auxiliary gearbox (see Section 10.3.10).

Mechanically driven air compressors were common on EE export locomotives, and given their associated simplicity will have been cheaper than the electrically-driven alternative. EE may have used the mechanically driven set-up to at least partially offset the high cost of the Napier engine in an attempt to achieve better price alignment with their competitors producing more conventional Type 2 designs.

The air compressor provided compressed air for the control of engine speed and for the operation of the locomotive brakes, electro-pneumatic control equipment, horns and sand ejectors. Air was built up to the required pressure, and stored, in reservoirs situated beneath the locomotive underframe.

A compressor governor controlled the operation of the compressor via an unloading valve in order to maintain air pressure in the main reservoir between 85 and 100lb/in². It should be noted that the values of the main reservoir pressures is significantly lower compared with today (118-140lb/in²), due to the fact that in the 1950s/60s most trains were vacuum braked and the compressor only serviced the brakes and other air requirements of the locomotive itself.

3.11.4 Radiator Fan
The radiator fan was mounted in the roof of the locomotive and was driven via a vertical shaft with flexible couplings from the auxiliary gearbox and included a centrifugal clutch to limit the torque imposed on the fan and fan shaft. The fan drew cooling air through the two radiators provided for cooling the engine lubricating oil and coolant and expelled it through the locomotive roof.

3.12 Electrical Equipment
3.12.1 Electrical Machines
English Electric electrical equipment carried structured design codes which can be summarised using EE835/7A and E835/8D, the main generators recorded as used in the 'Baby Deltic' fleet, as examples:

- EE   Designing company (could alternatively be DK [Dick Kerr] or others).
- 835   *Frame Number*, basically the model. These were allocated sequentially, within different 'hundred' blocks for different types of machine (e.g. 5xx for axle-hung, nose-suspended traction motors, 6xx for bogie- or frame-mounted traction motors, 7xx for various auxiliary machines (e.g. exhausters), 8xx for main generators, 9xx for auxiliary generators, etc.).
- 7A   *Form Number*; letters denoted an electrical variant and numbers a mechanical variant (although the latter did not necessarily run in a simple numerical sequence).

The 'EE835/7A & 8D' suffixes were quoted in a listing of electrical equipment supplied by GEC Alsthom to Mike Hunt in 1996. '7A' suggested (up to) seven mechanical variations and a single electrical variation or upgrade from the original; '8D' suggests yet another mechanical alteration on top of the previous ones, combined with three further electrical variations. It is not known what the exact differences were between the EE835/7A and 8D variants and may have been quite minor.

In reality, most published documents indicate the model number for the 'Baby Deltic' main generator as EE835A. The absence of a number in the suffix suggests that there were, either no mechanical variations from the original, or that the model number quoted deliberately ignored any mechanical variations and reflected electrical variations only. Thus EE835/7A could effectively be reduced to EE835A.

Alternatively, EE835A *may* refer to the installed generators pre-refurbishment, and the 7A and 8D variants to installed generators post-refurbishment, although there is no official documentation to prove or disprove this suggestion.

3.12.2 Main Generator

The main generator frame was bolted to the phasing gear casing on the engine via a flanged adaptor ring, and, together with the top-mounted auxiliary generator, formed one constructional unit with the engine. This unit was mounted resiliently on the locomotive underframe in a manner such that any underframe flexing was not transmitted to the power group.

The main generator was driven by a coupling which permitted a limited tolerance in the alignment of the engine and generator.

The main generator was of conventional design with separate excitation; electrical output was varied by control of engine speed and main generator field excitation. Although rated at 1,100hp in the 'Baby Deltics' the main generator was designed for development up to 1,250hp.

At 1600rpm, the main generator used in the 'Baby Deltic' was the fastest-rotating d.c. generator used in any BR diesel-electric locomotive. By rights, it should have had duplex lap winding to improve commutation at that speed; the EE831 generators in the prototype 'DELTIC' certainly did, as indeed did the GEC WT880 generators used in the NBL Type 2 fleet (D6100-57). Whether the 'Baby Deltics' featured lap winding has not been determined.

The generator supplied current to the four axle-mounted traction motors. In addition to its normal function, a series winding on the generator armature enabled the generator to be motored on current supplied from the batteries, for engine starting purposes.

3.12.3 Auxiliary Generator

The auxiliary generator (type EE912B) was mounted on top of the main generator and was driven by an auxiliary shaft from the engine phasing gears, putting out a constant 110V by a carbon-pile regulator. The four-pole d.c. machine supplied current for battery charging, the control system, the vacuum-brake exhausters, train heating boiler controls, lighting and cab heating.

As already mentioned, the No.2 traction motor blower was mechanically driven from an extension of the auxiliary generator armature shaft.

The output of the auxiliary generator was fairly restricted. Given the amount of mechanically-driven equipment on the 'Baby Deltics', it would appear that the whole design was a process of minimising auxiliary electrical loadings in order to maximise the traction capability from the main generator.

3.12.4 Traction Motors

The traction motors were resiliently suspended from the bogie frames. Each traction motor drove its associated axle through gearing comprising of a pinion on the motor armature and a gearwheel on the axle.

Four six-pole, force-ventilated, axle-hung, nose-suspended traction

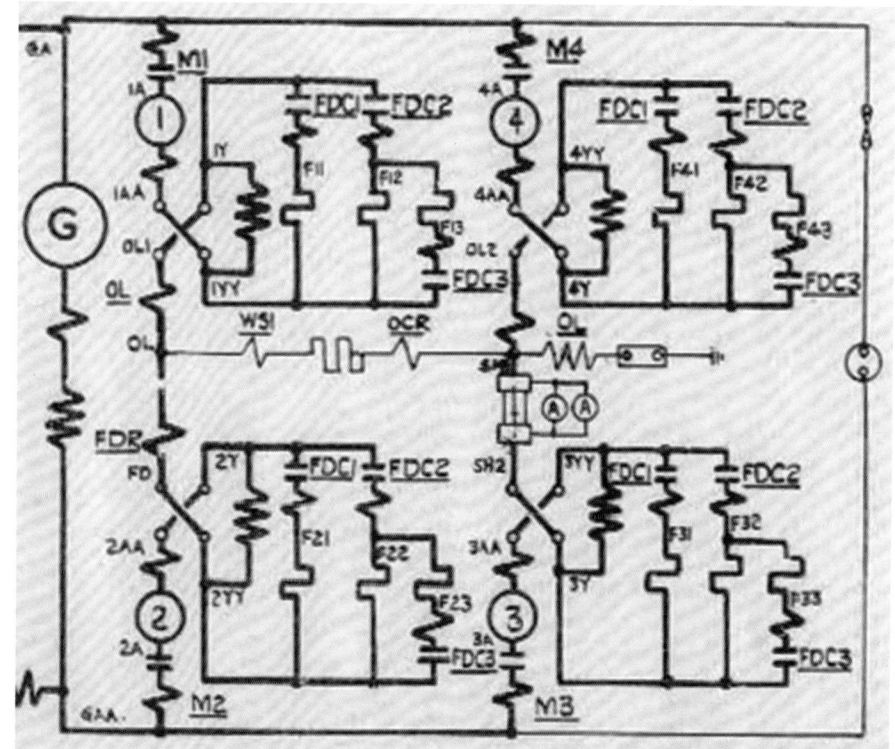

Four traction motor (M1 to M4) "series-parallel" configuration across the Main Generator (G).

FDC = Field Divert Contactors.

'Series-parallel' = two pairs of traction motors in series, each pair itself in parallel across the main generator.

The direction of locomotive movement was determined by the direction of current flow through the traction motor fields and was controlled by two reversing switches operated by means of a master switch in the driving cab. (Derived from EE Drawing RWS 81D-570 Part 2)

motors (type EE533A, one per axle) with a single reduction gear drive (63:17 ratio) were connected across the main generator in two parallel groups of two motors in series (Motors 1 and 2 in series, and Motors 3 and 4 in series, numbered from No.1 to No.2 end). This arrangement represented the so-called 'series-parallel' configuration.

The choice of motor connection, all-parallel or series-parallel, was a trade-off. When traction motors were connected in parallel, each motor essentially operated individually and was largely unaffected by its neighbours; so, if an axle slipped, the affected motor did not cause any collateral impact. However, the downside of all-parallel was that the generator had to be designed to deliver double the current at half the voltage compared with the series-parallel alternative. That meant large currents and significant amounts of copper, but less voltage and flashover risk.

With series-parallel configurations, the generator had to deliver twice the voltage (so that each motor still received the main generator output voltage of an all-parallel configuration) but half the current, so less copper and generally less weight, but with a greater propensity for wheel slippage and flashovers.

English Electric six-pole traction motors were used on their BR Type 1, 2 and 4 designs for BR (as well as 10203 and 'DELTIC'). These were heavier and more expensive than the earlier four-pole design on account of the two extra poles but provided a significantly smoother torque. However, they had a downside which was that their poles were physically closer together (i.e. 60° of arc apart, compared with 90° on the four-pole motors), meaning that if their commutators and/or brushgear were not in good condition, or the locomotive negotiated crossings and point-work at speed without any easing of power, or suffered severe wheel slippage, they were more prone to flashovers than the four-pole equivalents.

In order to extend the range of locomotive speed over which full engine power was available, provision was made for weakening the field strength of the traction motors by field divert resistances. As a locomotive accelerated, the traction motors generated an increasing 'Back EMF' (back electromotive force) opposing the flow of current from the main generator. This reduction in traction current would unload the generator (i.e. reduce its power), whose voltage would remain sensibly constant. However, unloading of the generator would be detected by the engine speed governor as an engine speed increase and the governor would respond to correct this and, in doing so, would also cause the load regulator to increase generator voltage to restore the engine/generator load balance. Ultimately, full generator voltage (set at the design stage and based on magnetic field saturation) would be reached.

At this point, the only way of preventing generator unloading was by reducing the adverse impact of the back EMF. This was achieved by field weakening, whereby switchgear operated a shunt across the field winding, so that some of the traction current was diverted away from the motor field system, but with all still flowing through the armature. The divert was a calculated resistance placed in parallel with the field. The sudden drop in traction motor field strength (and associated back EMF) caused a surge in current taken from the main generator, and this current enabled the locomotive to continue to accelerate. The field weakening process could take place several times, and in the case of the 'Baby Deltics', three field weakening stages were sufficient to enable the locomotive to reach their maximum line speed of 75mph. The three stages of field diversion were provided with the field strengths progressively reduced to 61.5%, 41.5% and 30.25% of full field.

Three Field Divert Contactors (FDC1 to FDC3 for each traction motor – see diagram on page 37) were energised automatically by the load regulator whenever the diesel engine was underloaded with the main generator fully excited. As each contactor was energised, its contacts closed, inserting equal resistances across the fields of all four traction motors simultaneously, causing the motors to speed up and hence increasing the locomotive torque/speed range.

In contrast, the Field Divert Relay (or, more accurately in terms of its function, the field *re*version relay) controlled traction motor field reversion to prevent the build-up of excessive currents in the motor circuits. If, when one or more stages of field diversion were in circuit, the current in the traction motor circuits exceeded a predetermined value, the series coil of the relay was energised and the field divert relay contacts were opened to progressively drop out the stages of field diversion in circuit at the time.

It is interesting to note that the traction motors originally specified for the 'Baby Deltics' were identical and, therefore, interchangeable with, the EE526 six-pole motors to be used on the Type A 1,000hp and Type C 2,000hp locomotives. However, in 1956 English Electric suggested the use of an alternative design (EE533A). Reference to this was made during a liaison meeting between BR and English Electric on 30 August 1956:

Minute No. 80. Traction Motors – Type 'B'
The Chairman referred to the EE Co's proposal to redesign slightly the traction motors in order to utilise more fully the rating obtainable from the generator and thereby securing a higher continuous tractive effort (TE) at lower speed. This will give a more desirable performance of the locomotive for the lower ranges of speed in freight train operation.

Previously the continuous TE was 25,000lb at 12mph whereas with the new arrangement 30,000lb continuous TE can be obtained at 10mph., the hp and overall efficiency of the locomotive remaining as originally submitted.

The effect of this change is that the traction motors of the Type 'B' (ER) and the Type 'A' (LMR) will not be interchangeable, but the advantages to be gained in securing a greater continuous TE at lower speed on Type 'B' mixed-traffic locomotives are, in the opinion of the CM&EE Eastern Region, of sufficient merit to warrant favourable consideration, and, provided the Commission has no objection to the proposal, the CM&EE is prepared to sacrifice interchangeability in the interests of high performance at lower speed. He has recommended the adoption of the EE Co. proposal to Mr. Warder [BTC Chief Electrical Engineer] from whom a reply is awaited and the Contractors will be advised of the result as soon as possible.

Permission was subsequently given by the BTC to use the EE533 motor instead of EE526. The loss of standardisation was considered to be less important than the improved traction torque/speed characteristics which suited the main duties intended for the 'Baby Deltic' fleet i.e. suburban passenger trains on the Great Northern main line out of King's Cross where 'fast getaways' between station stops was deemed critical. The bespoke motor was sufficiently different for a new Frame Code to be assigned rather than just a Form Number change to the EE526 Frame Code.

3.12.5 Exhausters
Vacuum for train braking was originally envisaged to be provided by two Northey-Boyce exhausters (type RE125) each driven by English Electric d.c. electric motors (type EE755/1A) but these were replaced, prior to delivery, by Reavell exhausters as part of overall locomotive weight reduction plans (see Section 5.3).

The replacement Reavell exhausters (5.25x10in), two per locomotive, were each driven instead by English Electric motors (type EE762A); under normal conditions the exhausters ran at 715rpm, although they could be speeded up to 1430rpm when rapid brake release was required. The motors were supplied with current by the auxiliary generator.

The 'surplus' Northey-Boyce exhausters and their associated EE motors were identical to those installed in the English Electric 2000hp Type 4 locomotives and it is believed that they found their way into these locomotives.

3.13 Train Heating Boiler
A Stone-Vapor (type OK4616) train-heating boiler of 1,600lb/hr capacity was fitted, housed behind the cab bulkhead at the No.1 end. Five hundred gallons of boiler water was carried.

At the time of order placement by BTC with English Electric in 1955, and indeed when the Purchase Order was placed by EE on Vulcan Foundry, the type of train heating boiler was not specified, only that a capacity of 1600lb/hr capacity was required. The Purchase Order stated that the boiler was 'to be of the Vapor-Clarkson or Clayton type' and that precise details would be settled later.

The following internal EE memos and letter from EE to the BTC describe how this subject developed:

Internal EE Memo 08/05/57
BTC Diesel-Electric Locomotive Type 'B' – Train Heating Boiler
We have just heard from Doncaster that Mr. Bond is unable to approve the fitting of Clayton boilers to these locomotives. Apparently he considers that further development work is necessary before it is acceptable. Doncaster have accordingly asked us to go ahead on the basis of the Vapor 1600lb/hr boiler.

Letter from EE to BTC (CM&EE), Doncaster, 19/12/57
Main-Line Diesel-Electric Locomotives Type 2 – Train Heating Boilers
Some time ago we sought your permission to fit these locomotives with Clayton boilers. At the time the trial Clayton boiler had not been fully tested by the London Midland Region and you ruled that Vapor Clarkson boilers were to be fitted.

We now understand that the London Midland Region has completed tests on the Clayton boiler and that a recommendation has been made to the BTC that it can be adopted as an alternative boiler on main-line locomotives. This being so we submit for your consideration a proposal that we should fit Clayton boilers to the last five locomotives.

The locomotive design is such that the Clayton or Vapor Clarkson boiler can be fitted with the minimum of alterations. We can also say that if you agree to the fitting of Clayton boilers… we can offer a reduction in price compared with the price of the Vapor Clarkson boilers.

Internal EE Memo 01/01/58
BTC Type 2 Locomotive (1,100hp) – Train Heating Boiler
We have been making one or two preliminary investigations into the position of the Clayton boiler and the following has come to light:

The boiler approved by the BTC after trials on a BTH locomotive is rated at 1,000lb/hr, whereas the one we shall have to use will be rated at 1600lb/hr. It would therefore seem somewhat doubtful whether Robert Stephenson & Hawthorn could produce a 1600lb/hr boiler in time for fitting on No.6 locomotive [D5905]. A further point is that the control gear for these boilers is being developed by TCD [Traction Control Design Department] so as to be more suitable for traction services. It is not expected that this development will be completed before June 1958.

The option of deploying Clayton boilers was never taken presumably on the basis that the intended July 1958 introduction for D5900 would have precluded sufficient development time for the higher-rated Clayton boiler.

3.14 Bogies

The fabricated bogies were very similar to those used on the EE 1,000hp Type 1 locomotives with 3ft 7in diameter wheels and an 8ft 6in wheelbase. The bogies utilised welded and riveted side-frames of box section with welded cross members riveted to the sides. The bolster was carried on semi-elliptical springs supported by planks and inclined swing links; lateral movement of the bolster was limited by rubber blocks. Springing between frame and wheels was in the form of nests of helical springs located between the frame and an equalising beam on each side, the beams being underslung from SKF (Skefko) roller-bearing axle-boxes which had the usual guides and manganese-steel liner plates. Similar plates took up the traction and braking forces between bolster and bogie frame.

Westinghouse straight air brakes applied two blocks on every wheel through clasp rigging, actuated by individual cylinders for each wheel.

Following reports of poor riding qualities by BR on the EE Type 1 locomotives, BR questioned the use of the same bogie on the 'Baby-Deltics'. In the event, BR reports proved to be somewhat exaggerated and the original bogie design was retained. English Electric, however, looked at fitting shock absorbers or dampers between the equalising beams and the bogie frame. Whilst not fitted on delivery, all locomotives were fitted with such equipment later in life.

As regards bogie dampers, the following internal EE memo sheds some further light on the subject:

Internal EE Memo, 30/09/59
BTC Type 2 DE Locomotives – Hydraulic Dampers
Provision was originally made for the fitting of hydraulic dampers in the primary springing of these locomotives, but these were subsequently removed as part of the weight reduction modifications [prior to delivery]. We have now been asked by Mr. Lyons (RSDD, Preston) to obtain customer's permission to refit these dampers … at Hornsey Motive Power Depot.

Customer has now given his approval to this proposal and we shall be obliged if you will arrange to have the dampers refitted, bearing in mind customer's request that this work shall be done on examination days, so as not to necessitate locomotives being withdrawn specifically from traffic for this purpose.

An internal EE memo, dated 8 January 1965, revealed that the hydraulic dampers 'were fitted to nine of the ten locomotives after delivery'. Photographic evidence indicates that D5903 was the locomotive not so fitted, although it did ultimately receive dampers as part of the rehabilitation process in 1964.

3.15 Braking

Locomotive braking was by air, being proportionally controlled by the vacuum system when working a train. Westinghouse equipment was deployed.

An interesting feature of the brake system was the provision of 'trip cock' equipment, a requirement for locomotives which operated over the London Underground Metropolitan 'Widened Lines' between King's Cross and Moorgate shared with London's Underground network. Modified vacuum piping and valves, and trip-cock apparatus were fitted to the bogies and buffer beams during commissioning at Doncaster Works (see Section 16.1.5).

3.16 Control System

A fundamental part of diesel-electric locomotive operation is the control system. All such control systems feature, at their core, a diesel engine speed governor, deployed in conjunction with a torque (load) regulator which controls generator output.

The system employed on the 'Baby Deltic' locomotives was broadly similar to that used on the EE Type 4 locomotives (Class 40); however, the engine speed governor used in the 'Baby Deltics' was of Ardleigh Engineering manufacture, as opposed to English Electric's own design of servo-governor. The extremely sensitive Ardleigh was more suitable for quick-running engines such as the Napier T9-29.

The governor regulated the idling speed of 600rpm, and set and controlled the engine speed at any desired value between idling and its upper speed of 1,600rpm. The governor was mounted on top of the engine block over the 'AB' crankcase, the drive being taken from a gear train from the 'AB' crankshaft phasing gear. The governor incorporated an oil pump, oil pressure being the medium used to operate the governor servo. The required speed settings were impressed on

D5909 No. 1 end bogie, 30A Stratford, 3 January 1972. Nine holes drilled out of the equalising beam. Shock absorbers fitted 'outside' of the two helical springs between the bogie side frame and the equalising beam. All sandboxes removed. (John Grey Turner)

D8005, 1A Willesden, 25 March 1961. Apart from the cab access step arrangements and external pipe runs, this picture provides an illustration of how the 'Baby Deltic' bogies would have looked but for the need for major weight-savings initiatives. Oblong cross-section equalising beams, four sandboxes per bogie (two each side) and oval buffers. (Rail-Online)

the governor mechanism by an air-operated actuator in accordance with the position of the control handle in the cab.

The torque (or load) regulator used in the 'Baby Deltics' was type EE KV5-A3, part of the older KV5 family of load regulators in which the drive to the rotor arm was by means of an electric pilot motor. A major difference between this type of regulator and that used in the Class 40 was that the pilot motor which powered it was reversed by changing the direction of current flow in the motor armature as opposed to changing the polarity of the poles. The operation of the torque regulator motor was controlled automatically by two cam switches actuated by the engine governor.

The torque regulator comprised a row of fixed contacts arranged concentrically around a contact roller. As the arm rotated, the roller opened and closed the fixed contacts, or 'finger switches', and, in doing so, controlled the strength of main generator excitation and the introduction of traction motor field diversion.

In very simple terms, the governor controlled the quantity of fuel injected into each cylinder immediately before the start of the power stroke. Provided there was sufficient air for combustion, the quantity of fuel injected determined the torque (or turning force) produced by the engine. The torque was proportional to the position of the rack which controlled each fuel pump and, therefore, the quantity of fuel injected.

The principal function of the governor/torque regulator system working in conjunction, was to set the position of the fuel racks correctly for any desired engine output and maintain it and the matching engine speed for that selected output, against changes in electrical loading.

For any driver's control handle position, the torque regulator ran up until the engine was fully loaded for that specific engine speed. As the control handle was advanced between 20° and 90° from 'OFF', the engine governor speeder linkage caused the torque regulator to run in the direction to increase the main generator field, and simultaneously increasing engine speed. Power output was therefore progressively increased as the control handle was advanced. Engine speed setting was achieved pneumatically in accordance with the 'Blue Star' scheme for multiple-unit control, and was stepless and continuous. In conjunction with this, three stages of traction motor field weakening were employed (see Section 3.12.4).

The 'Deltic' engines had no flywheel to store kinetic energy, as their triangular configuration did not lend itself to the inclusion of such a device. As a consequence they had a very low moment of inertia, and so protection against under- and over-speeding in the event of sudden loading or rejection of load was vital. Despite the Ardleigh governor's sensitivity, an under-speed device linked electrically to the main generator excitation automatically unloaded the engine when the crankshaft speed fell below 530rpm. An overspeed device linked mechanically to the fuel injection pumps control shaft automatically stopped the engine if the engine speed attained 1,800rpm. It was possible that the under-speed device on the 'Baby Deltics' was isolated at some stage; it certainly was on the production 'Deltic' Type 5s as it tended to cause 'inappropriate' engine shutdowns on occasions.

All of the control gear on the 'Baby Deltics', with the exception of the two master controllers in the driving cabs, was contained in one cubicle in the engine room, which formed part of the No. 2 cab rear bulkhead. The master controllers were type EE KM26 (specific variant not known).

The 'Baby Deltics' could be operated in multiple with other Types 1, 3 and 4 locomotives built by English Electric, together with other locomotives fitted with the electro-pneumatic ('Blue Star') control system, requiring electrical control cables and engine control air pipes between locomotives.

The control system included a pedal-controlled deadman system. Wheel slip protection circuits were included, and a visual indication of slip was given to the driver.

An overload and earth fault relay was connected in the traction motor circuits in such a way that, if the current in the circuit to either pair of traction motors reached a dangerously high value, or if a power-to-earth fault developed, the relay tripped and its contacts were opened to de-energise the power control relay, thus causing instantaneous power cut-off and simultaneously returning the engine to idling speed.

Chapter 4
APPEARANCE DESIGN AND STYLING

There is surprisingly little information regard the appearance work undertaken on the 'Baby Deltics'.

Professor R.D. Russell (of R. D. Russell & Partners, London) was the nominated Industrial Design Consultant for a series of English Electric locomotives i.e. Type 1s (D8000), 'Baby Deltic' Type 2s and Type 4s (D200), explaining the similarity of the latter two classes and also their common features with some of the earlier EE diesel export designs.

Whilst Russell dealt with the physical design aspects, as will be seen in Section 17.1, E.G.M. Wilkes (of Wilkes & Ashmore, Horsham) dealt with the more detailed aspects of the locomotive design including the livery.

Given below are two comments regarding the 'Baby Deltic' physical design :

"A good-looking locomotive… with rather stubby nose units…" (Brian Webb, 1976).

The last of the Type 2 diesels to appear in service, the English Electric 'Small Deltic', is a very interesting machine which differs from the other manufacturers' Type 2 designs in having a prominent nose at each end of the locomotive to house the vestibule gangways, a feature previously found only on the larger Type 4 diesels. The design of the driving cab is a slightly altered version of that used on the D200 class, with a somewhat shortened nose end, and together with the whole of the superstructure is neat and clean in appearance. (*Trains Illustrated*, September 1959).

Following representations by the BR Eastern Region and English Electric, a decision was made by R.C. Bond (BTC CME) during July/August 1957 to fit windows in the sides of the 1,100hp 'Baby Deltic' and 2000hp locomotives, four each side, in order to give daylight within the locomotive body for footplate and maintenance staff. Preliminary drawings had already been produced by English Electric to illustrate their proposals regarding the shape and disposition of the windows and Bond suggested that these should be considered by Wilkes within the context of the overall design; however, to avoid any delays to production, English Electric insisted on the original drawings being accepted and these were indeed agreed by Bond on 8 August 1957.

D5900, Doncaster Works (Paint Shop Yard), Undated. (Geoff Sharpe)

Chapter 5
DELIVERY AND ACCEPTANCE

5.1 Introduction to Traffic

Ordered	16/11/55
Original Delivery Promise (First locomotive)	12/07/58
First Locomotive Delivered	17/04/59

The delivery of the first locomotive was based on 30 months following BTC order placement subject to resolution of technical issues (achieved January 1956).

The English Electric Purchase Order placed on Vulcan Foundry for the construction of the 'Baby Deltics' dated 23 January 1957 included contractual delivery dates ('delivered free on rail at our Works' - see table below).

The BTC's General Conditions of Contract for Locomotives and Rolling Stock provided for the payment of liquidated damages at the rate of half of one per cent of the contract unit price per week, subject to a maximum of 15 per cent, if the contractor failed to make delivery at the proper time and, as a consequence, the commission suffered loss. The contractor was entitled to ask for an extension of time for completion if during the performance of the contract it appeared that a delay was likely to arise because of industrial disputes or other causes outside the contractor's control.

In the event, the first 'Baby Deltic', D5903, was delivered nine months later than the date when D5900 was supposed to have been delivered and was among the last of the 'Pilot Scheme' locomotives delivered to BR. As can be seen from the table below, all of the 'Baby Deltics' triggered the late delivery penalty by a considerable

Contractual Delivery Dates.

Loco No.	Contractual Delivery Date	Actual Delivery Date	Weeks Late	Indicative Penalty
D5900	12/07/58	22/05/59	45	15.0%
D5901	30/09/58	22/05/59	33	15.0%
D5902	15/10/58	01/05/59	28	14.0%
D5903	31/10/58	17/04/59	24	12.0%
D5904	15/11/58	24/04/59	23	11.5%
D5905	30/11/58	08/05/59	23	11.5%
D5906	15/12/58	08/05/59	21	10.5%
D5907	31/12/58	15/05/59	20	10.0%
D5908	15/01/59	29/05/59	18	9.0%
D5909	31/01/59	19/06/59	20	10.0%

margin, although whether the BTC actually applied the penalty payments is unknown.

5.2 Reasons for Delays

The smaller T9 'Deltic' power unit used in the 'Baby Deltics' was not downrated as much as D18 engines used in the 'DELTIC' prototype, and at 1,100hp at 1600rpm it exhibited a relatively high output level. Due to the high output speed, some innovations inevitably had to be made in terms of gearing and output drive to a number of otherwise standard items of auxiliary equipment (such as cooling fans and air compressors) that had been designed to work with slower, four-stroke diesel engines. These fairly complex changes took significantly longer to develop than expected.

The most significant factor causing delay was the weight of the locomotive which far exceeded the axle-loadings required by BR and as a consequence the planned delivery dates of between 12 July 1958 and 31 January 1959 had to be abandoned. This topic is dealt with in the following sections.

5.3 Weight Issues – A Type 2 Problem

Achieving the power and axle-weight specifications for the Type B Pilot Scheme locomotives proved a difficult issue for all of the manufacturers involved in producing the BTC Pilot Scheme locomotives either at the design stage, or, as we will see later, at the time of their introduction into traffic.

The BTC imposed stringent weight limits on the Pilot Scheme Type 1 and Type 2 designs to match branch line civil engineering limitations; thus the maximum axle loading of 18 tons (and 17 tons for the Western Region) was specified for both types. The specification with respect to the Type 1s, largely intended for freight duties, was easier to achieve, given the avoidance of certain items of equipment such as a train heating boiler.

English Electric opted for the Napier T9 engine with the obvious benefit of being both a high-speed and lightweight power unit which, therefore, had the potential to resolve the power/weight requirements. The NBL Type B locomotives for the Western Region (D6300) also deployed high-speed MAN-designed engines alongside hydraulic transmissions; the equivalent diesel-electric locomotives for the Eastern Region (D6100) utilised the same MAN high-speed power unit.

Brush/Mirrlees (D5500 locomotives), Metropolitan-Vickers/Crossley (D5700), BR/Sulzer (D5000) and Birmingham Railway Carriage & Wagon/Sulzer (D5300) opted for the traditional heavy-weight four-stroke diesel engines. Such was the weight of these engines that Brush and Metropolitan-Vickers had to deploy 6-axle (A1A-A1A) and 5-axle (Co-Bo) bogies respectively to achieve the required axle-loads.

Despite best endeavours, the English Electric, NBL, BRCW and BR designs all ended up supplying overweight locomotives at the time of their introduction, which necessitated a degree of re-engineering to resolve the problem, a revision to the duties carried out by the locomotives, and a degree of relaxation by the BR civil engineers with respect to axle loading.

The English Electric 'Baby Deltic' design, despite the lightweight engine, did not preclude these locomotives from being overweight at introduction. It has been argued that the 'Baby Deltics' would have been too light had it been designed 'as normal' and would have struggled with adhesion issues, hence the suggestion that the superstructure was deliberately constructed with heavier materials in order to offset the light weight of the power unit. Ultimately 'over-enthusiasm at English Electric' (Brian Webb, 1976) resulted in them overshooting the mark, not helped by some auxiliary components being heavier than prescribed by suppliers at the design specification stage.

5.4 Weight Resolution

In early 1958 the weighing of the first 'Baby Deltic' at Vulcan Foundry showed that the specified weight of 72tons had been exceeded by 3tons. This was not the first time that this problem had occurred as the BRCW/Sulzer Type 2s (D5300-19) also destined for Hornsey were also found to be overweight; this class was accepted into traffic from mid-1958 at 77tons 17cwt and, as a direct consequence, were immediately precluded from transfer freights to the Southern Region.

The excess weight of the 'Baby Deltics' was totally unacceptable to BR so EE had to do some serious thinking to achieve the required weight reduction. BR specifically focussed their attention on the EE Type 2s rather than both Classes because by 1958 it had already been planned for the BRCW and NBL locomotives to be moved away from the Eastern Region once the Brush/Mirrlees Types had been delivered in sufficient numbers as part of the BTC class concentration initiative. Transfer of the BRCW and NBL Type 2s to the Scottish Region commenced in April 1960.

The excess weight of the 'Baby Deltics' was as a consequence of several factors:

- over-specification of certain mechanical parts to partially mitigate against the light weight of the Deltic diesel engine in order to achieve the 72ton overall weight to maximise adhesion,
- heavier than specified auxiliary equipment from external contractors (train heating boiler, compressor, brakes, fire fighting equipment, AWS equipment, flexible gangway connections, etc.), and,
- the provision of radiators capable of supporting 1375hp engines.

A major example of an overweight auxiliary was the OK4616 train-heating boiler which should have weighed 2750lb according to Vapor-Clarkson in the USA. EE worked to this weight during the locomotive design stage but when the British-built Stone-Vapor boiler was delivered, with standard BR fittings, it weighed 3448lb, 700lb more than originally specified. EE believed that this was going to cause them 'serious embarrassment' given the need to approach BR for a weight concessions in water and fuel tank capacity, or, permission to deploy a lower capacity boiler (e.g. OK4610 model, 1,000lb/hr).

Subsequent joint EE/Stone discussions identified a weight reduction of 260lb achieved by actions including the re-design of boiler fittings and mountings, the substitution of the steel boiler cabinet doors and side panels with glass-reinforced plastic, and smaller filters.

EE highlighted the issue with BTC with a view to some sort of concession, but the BTC were unwilling to compromise as the following letter illustrates:

Letter from BTC(CM&EE), Doncaster to EE, Bradford, 3 April 1958
Main-Line Diesel-Electric Locomotives Type 2 - Stone Vapor Steam Generator
If you are still unable to come within the 18ton axle load I would be prepared, as a last resort, to concede a little water capacity, say 50 gallons maximum, but I am reluctant to do this as the 600 gallons provided is none too much during the heating season. Apart from this I should in any case receive strong criticism from the operating interests.

To use the modern vernacular, serious amounts of brainstorming were undertaken, as a consequence of which the following actions were identified to lighten the overall locomotive weight:

- drilling of circular holes into the bogie frames and equalising beams;
- replacement of standard oval steel buffers by circular Oleo-pneumatic buffers;
- substitution of steel buffer blocks with aluminium;
- substitution of the removable steel roof sections with aluminium alloy;
- replacement of steel fuel and water tanks with alloy;
- deployment of aluminium radiator shutters;
- removal of the inner sand boxes and replacement of the remaining four by aluminium in place of steel;
- omission of the flexible gangway connections (with BR's permission).
- replacement of the Northey exhausters (and associated EE755/1A motors) by Reavell units (with EE762A motors).

In the event, the aluminium sandboxes were not fitted until the early 1960s.

To minimise costs, a number of the redundant items were subsequently used by English Electric in their 1,000hp and 2,000hp locomotives.

In January 1959, prior to applying the various weight-saving initiatives, EE made another approach to the BTC for concessions but, once again, the response was not positive from EE's perspective:

Internal EE memo from RSDD, Preston, to Traction Department, Bradford, 12 January 1959
Type 2 Diesel-Electric Locomotives
On 2nd January we discussed with Mr. J.F. Harrison the weight of these locomotives, and later indicated that we wished to supply them at a weight of 74.5tons with

two-thirds water and fuel and sandboxes full.

We now have a letter from Mr. Harrison stating that at this weight it would be impossible to obtain any acceptable utilisation from them, and suggesting that we should arrange with Mr. E.S. Cox for a meeting to discuss in detail what would be involved in further weight reductions to give the maximum axle loads of 18tons.

This reaction triggered application of the various modifications. Interestingly, the drawings for the Oleo pneumatic buffers were only signed-off by English Electric in February 1959 and deliveries of the Reavell exhausters did not commence until mid-March.

As a reference starting point, D5900 was weighed on 23 January 1959 with the following results:

Water/Fuel Tanks
   and sandboxes
   two-thirds full   74t 7cwt 1qr
Full services
   (actual)   75t 13cwt 0qr

Following modifications on D5901, English Electric undertook a number of re-weighing exercises at Vulcan Foundry during March 1959. It would appear that EE were not entirely convinced that the various weight-saving initiatives would be sufficient and extreme lengths were taken by the Chief Draughtsman to achieve the lowest overall weight possible and balanced axle-loadings, including:

Preparation of track. The rails on which the locomotive was weighed were to be perfectly level and any packings required were to be placed immediately below the axle centres to prevent deflection of the rails when the locomotive was in position.

Preparations for locomotive weighing
- Engine oil and coolant water filled up to two-thirds of the max-min running level in the tank.
- Fuel oil tank contents (engine tank only) at various prescribed levels.
- Water tank contents at various prescribed levels.
- Northey exhausters and associated gear removed and weights of 11cwt weight placed in position of removed equipment.
- All batteries in original fuel tank position, or, removed and 7cwt weights placed in position in the passageway near the train heating unit, between the cab door and the radiator.
- Ballast: Arrangement(s) to be made to place up to 6cwts of weight at the extremity of No.1 end.
- Four sandboxes: contents to be at various prescribed levels.
- Buffers: Weights of 175lb each to be placed in position to represent the Oleo buffers.
- Bogies to be fitted with 5¼in wide tyres.
- All modifications as called for up to date.

After each weighing the scales were to be transferred end for end to give an accurate check.

The results of weighings of D5901 undertaken on 12 and 13 March are listed below:

BTC CCF0875 1,100hp Diesel-Electric Locomotive (Type 2) Locomotive No. D5901
Weight Distribution Diagram – Weighed 12/03/59 at Vulcan Foundry (average of two weighings)
Axle 1: 17-19-2   Axle 2: 17-16-2
Axle 3: 18- 4-1   Axle 4: 18- 3-0
Total: 72- 3-1

Fuel (engine)	250gals (half full)
Fuel (boiler)	Nil
Water (boiler)	200gals (half full)
Lub Oil	Two-thirds of max/min. running level
Coolant	Two-thirds of max/min. running level
Sand	240lb
Exhausters	11cwts
Buffers	6¼cwts

ATC reservoir near boiler.

Weight Distribution Diagram – Weighed 13/03/59 at Vulcan Foundry (average of two weighings)
Axle 1: 18-14-0   Axle 2: 18-10-1
Axle 3: 18-10-2   Axle 4: 18-10-2
Total: 74- 5-1

Fuel (engine)	450gals (full)
Fuel (boiler)	Nil
Water (boiler)	500gals (full)
Lub Oil	Two-thirds of max/min. running level
Coolant	Two-thirds of max/min. running level
Sand	360lb

Batteries moved from fuel tank, 7½cwts placed near to boiler 'A' side to compensate

Exhausters	11cwts
Buffers	6¼cwts

ATC reservoir near boiler.

Weight Distribution Diagram - Weighed 13/03/59 at Vulcan Foundry (average of two weighings)
Axle 1: 18- 5-3   Axle 2: 18- 3-3
Axle 3: 18- 5-2   Axle 4: 18- 3-3
Total: 72-18-3

Fuel (engine)	300gals (two-thirds full)
Fuel (boiler)	Nil
Water (boiler)	330gals (two-thirds full)
Lub Oil	Two-thirds of max/min. running level
Coolant	Two-thirds of max/min. running level
Sand	240lb

Batteries moved from fuel tank, 7½cwts placed near to boiler 'A' side to compensate

Exhausters	11cwts
Buffers	6¼cwts

ATC reservoir near boiler.

Weight Distribution Diagram – Weighed 13/03/59 at Vulcan Foundry (average of two weighings)
Axle 1: 18- 1-3   Axle 2: 17-19-0
Axle 3: 18- 2-1   Axle 4: 18- 1-0
Total: 72-4-0

Fuel (engine)	250gals (half full)
Fuel (boiler)	Nil
Water (boiler)	200gals (half full)
Lub Oil	Two-thirds of max/min. running level
Coolant	Two-thirds of max/min. running level
Sand	240lb

Batteries moved from fuel tank, 7½cwts placed near to boiler 'A' side to compensate

Exhausters	11cwts
Buffers	6¼cwts

ATC reservoir near boiler.

Despite the severe restrictions of in-service materials (fuel/oil/water/sand) D5901 was still above the required 72tons in all instances, the closest at 72tons 4cwt including unrealistically low liquid levels.

On 7 April D5903 (not D5909 as frequently quoted) was sent to BR Doncaster Works for 'official' weighing at 10.00a.m. the following day with EE personnel in attendance. The weighings were again based around a number of prescribed conditions as follows:

- Engine oil filled up to two-thirds of the max-min running level in the tank.
- Coolant water filled up to two-thirds of the max-min running level in the tank.
- Fuel Tank (Engine). Various prescribed levels.
- Fuel Tank (Heating Boiler). Completely empty.
- Water Tank. Various prescribed levels.
- Northey Exhausters and associated equipment removed. Reavell exhausters fitted.
- Air reservoirs and frame. Modifications completed.
- Batteries in original position, half in fuel tank recess and half in water tank recess.
- Sandboxes. Two-thirds full (4 at 60lbs each).
- Oleo buffers fitted.
- The locomotive to be complete with all details.

Variations in the amounts of liquids and sand were set up and the results achieved are provided below.

D5903 - Record of Weights taken at Doncaster Works on the Voiron Weigh Table, 8 April 1959

D5903, returned to Vulcan Foundry on 9 April. Inevitably there was some argument about the accuracy of Doncaster's machine, but BR demanded that a further 6cwt reduction was necessary to achieve the required 72ton overall weight (albeit with limited fuel and water provision).

D5903 – Record of Weights.

Weigh	Fuel (gals)	Water (gals)	Sand (lbs)	Axle 1 T-C-Q	Axle 2 T-C-Q	Axle 3 T-C-Q	Axle 4 T-C-Q	Total T-C-Q
1st weigh	270	250	240	18- 1-2	17-17-3	18- 2-1	18- 4-3	72- 6-1
2nd weigh	230	300	240	18- 2-0	17-16-3	18- 3-0	18- 4-3	72- 6-2
3rd weigh	250	250	240	18- 1-0	17-17-0	18- 1-3	18- 4-2	72- 4-1
4th weigh	350	450	240	18- 8-0	18- 4-0	18- 5-2	18-11-0	73- 8-2
5th weigh	450	500	360	18-10-1	18- 4-3	18-11-0	18-11-0	73-17-0

To achieve this, EE put in hand the following modifications:

- Remove remainder of redundant gangway items.
- Cut holes in engine room floor under main generator and exhauster and cover with thinner gauge steel sheet.
- Replace steel roof above control cubicle with aluminium.
- Fit lighter reservoir mountings.
- Replace three steel reservoirs with ⅛in thick stainless steel ones.

These changes resulted in a weight reduction of 1165lb, and a revised locomotive weight of 71tons 16cwt 12lb. Further weight reduction was not considered necessary although an internal EE document dated 23 April exists which included the following additional actions (although whether these were undertaken is not known):

- Fuel tank. Aluminium instead of steel, moved forward and boiler section deleted.
- Fuel filling and drain cocks. Moved to the position of the original boiler fuel units.
- Batteries. Moved forward in the fuel tank.
- Air Reservoirs. Mounted as close as possible to the new fuel tank.
- Reservoir mounting frame. Deleted and new separate mounting details direct to the frame.

The documents listed the revised calculated weights as: No.1 bogie 35.724tons; No.2 bogie 35.800tons

A number of observations need to be made:

- What was the point of achieving the 72tons overall weight when the locomotive was effectively not in an everyday operational condition (regarding fuel and water consumption requirements)?
- The official BR weight drawings for the 'Baby Deltics' show an overall weight of 73tons 17cwt, the weight of the locomotive with full service provision as per the fifth weighing on 8 April i.e. nearly two tons overweight.
- Certain pieces of ejected equipment were re-installed prior to entry into traffic e.g. gangway door equipment, train heating boiler panelling, etc.
- As far as is known the re-positioning of the batteries was never undertaken and presumably allowing the unequal axle-loading to continue.
- Most importantly, the weight reductions achieved still precluded their use on cross-London traffic to the Southern Region and in the short term steam locomotives were retained to cover this traffic until 1962.

The full list of actions to achieve the weight-reduction modifications to meet the BR axle-load limitations had been identified by April 1959. By this time, the construction of the ten locomotives on order was at an advanced stage and ostensibly ready for delivery; however, the identified weight-saving modifications had to be retro-fitted which inevitably led to both further delivery delays and the introduction of locomotives out of numerical sequence.

5.5 EE Test Runs

D5902 undertook two test runs in early April 1959, the

D5902, Garstang, 3 April 1959. Discs set to the express passenger position. Load trial between Newton-le-Willows and Penrith via Edge Hill (and return). On this trial, BR drivers, inspectors, VF staff, EE and Napier engineers all crowded into the locomotive; the lack of gangway connections at that time precluded any access between the locomotive and train whilst in motion. (Rail-Photoprints)

first from Vulcan Foundry, Newton-le-Willows, to Chester and return, and the second from Newton-le-Willows to Penrith via Liverpool Edge Hill and Preston. These trials were sponsored by English Electric and were in effect running-in trials rather than BR acceptance trials.

An internal English Electric memo, dated 18 March 1959 provided a considerable amount of detail regarding the forthcoming trials:

Memo from Chief Electrical & Test Engineer (Traction) to Chief Asst. Works Manager, 18 March 1959
BTC Type 2 Locomotives – Test Run

Subject to confirmation from BR(LMR) the run to Chester and back with locomotive D5902 and passenger brake will be made on Wednesday, 1st April. The loaded run to Penrith and back via Edge Hill with locomotive D5902 and eight coaches [plus passenger brake] will be made on Friday, 3rd April.

As far as we can ascertain, the timings for these runs will be the same as for the previous runs carried out with the Type 1 locomotives. These were:

Vulcan			dep.	11.55a.m.
Chester	arr.	12.44p.m.	dep.	2.30p.m.
Vulcan	arr.	3.23p.m.		
Vulcan			dep.	8.05a.m.
Edge Hill	arr.	8.50a.m.	dep.	9.40a.m.
Penrith	arr.	12.22p.m.	dep.	1.00p.m.
Edge Hill	arr.	3.19p.m.	dep.	3.39p.m.
Vulcan	arr.	4.14p.m.		

As these runs are really Works Trials there will be quite a lot of special tests being carried out in the engine room with personnel from TCDD, Bradford, and our Loco Test Department. This will limit the number of passengers who can be accommodated in the locomotive especially on the Penrith run. Provisionally, the number of people engaged on the tests from the purely practical angle will be five, comprising three from the Loco Test Department and two from TCDD, Bradford. It is probable that a Napier representative from Controls Section, Acton, will be present to observe the engine governing in conjunction with the torque control scheme on the Penrith run. It will be necessary to take along one Fitter and one Electrician as passengers in case of complications.

It is thought worthwhile to raise the issue of passage between locomotive and train now that the corridor connections have been removed. Will it be possible to use this means of access in service? If not, would it not be prudent to provide means of preventing someone from inadvertently opening the doors?

A memo on procedure to be adopted for these test runs will be issued after a meeting has been held with the TCDD, Bradford, to discuss the tests and observations required. A list of tools and equipment to be carried is also being prepared.

On the Chester run, the following maximum speed restrictions were applied: Newton-le-Willows to Warrington: 40mph; Warrington to Frodsham: 60mph, Frodsham to Chester: 75mph.

Extracts from the two test run reports with D5902 are given below:

Test Report on Works Test Run to Chester, 1 April 1959
Newton-le-Willows to Chester via Frodsham. Load: Passenger brake.
Departed Works: 1157hrs, Returned to Works: 1430hrs.

General Remarks: Three times during the run the engine overspeed governor operated when shutting off smartly on the controller. Each time the governor was re-set without hindrance to the trial. In return to the Works the setting of this overspeed governor was found to be 1730rpm which is almost equal to the engine no-load speed. Napier's Mr. Yates arrived on Thursday, 2nd instant, and re-set this to operate at 1800rpm. Loco riding is exceptionally good.

Conclusion: A most satisfactory run which brought some points to light such as the o/s governor and wheelslip equaliser circuit which could not be checked on static tests. Vibration of cooker and supports was very bad and will undoubtedly show up even more with the new modified cooker supports as fitted to loco D5903.

D5902 Test Report on Works Test Run to Penrith, 3 April 1959
Newton-le-Willows to Penrith via Edge Hill and Preston.
Departed Works: 0805hrs, Returned to Works: 1600hrs.
Load: Passenger brake, then 9 coaches, trailing weight 276tons from Edge Hill.

General Remarks: The run was satisfactory in all respects with the exception of the ascent of Shap. Despite a request to Preston Control we were not given a clear run up and therefore it was impossible to check FDR relay and Field Reversion.

On Wheelslip Operation there is overshoot on the torque regulator by a few contacts but this did not cause any difficulties to arise. An excellent example of wheelslip operation occurred on starting the train at Scout Green Box on wet rail when the driver notched up too far on the controller. The wheelslip system was allowed to function without moving the controller and power was picked up again and the train pulled away without further difficulty. Sand was applied for about 3 seconds only.

There were messy deposits of oil from the engine manifold drains which drip on the engine and walk-way. These were pointed out to Napier's Service Dept. Rep. in October 1958 but no action has yet been taken to pipe these down to the engine well. Vibration of blower duct at No.1 end very bad (Preston RSDD informed about this over 3 months ago). Riding of loco is very good, being vastly superior to Type 1 loco.

Conclusion: As far as the data contained in this report shows, the locomotive is satisfactory but complete satisfaction must be withheld until field reversion stages and associated torque regulator behaviour have been observed on the next load run.

A post-trial internal EE Memo provides some further insights:

Memo from TCDD to Chief Engineer, Vulcan Foundry, 8 April 1959
BTC Type 2 Locomotive – Report on Works Trial Run: Loco D5902
We would like to record our appreciation of the work done by Vulcan Foundry in carrying out these trials last week and in particular of Mr. Cameron's report about them received here today. Incidentally, it shows the value of trials carried out without Railways' influence.

We have one comment on the trials extra to Mr. Cameron's with which we fully agree. Engine noise in No.1 cab appears as a high-pitched whine presumably from the engine air-charger. It will cause complaints."

5.6 Doncaster Works Acceptance Trials

In the BR system for testing privately-built locomotives, responsibility for the testing of completed locomotive prior to acceptance into traffic was with the Chief Mechanical & Electrical Engineer's department of the sponsoring Region. In the case of the 'Baby Deltics', the sponsoring Region was the Eastern Region, with the Mechanical Engineer (Design) at Doncaster having ultimate responsibility.

Delivery from Vulcan Foundry to Doncaster was undertaken by a BR driver in conjunction with English Electric personnel. EE engineers were available at Doncaster up to the time of acceptance, including during static and main-line pre-acceptance trials.

D5906 and D5905, Doncaster Works (Paint Shop Yard), May 1959. NBL, BRCW and Brush Type 2s for company. (Nigel Petre)

Details of main-line acceptance trials undertaken by the 'Baby Deltics' are given below (Sources: BLS *Railway Locomotives* and CD&E6):

Acceptance Trials.

Loco No.	Dates ex-EE/VF	Test Train Dates	Comments
D5900	22/05/59	26/05/59	Test-run in multiple with D5901
			To Hornsey 08/06/59 (with D5905)
D5901	22/05/59	26/05/59	Test-run in multiple with D5900
D5902	01/05/59	06/05/59	Noted at Doncaster Works 10/05/59
			To Hornsey 12/05/59 (with D6103)
D5903	17/04/59	21/04/59	To Hornsey 24/04/59
D5904	24/04/59	29/04/59	To Hornsey 01/05/59
D5905	08/05/59	12/05/59	Noted at Doncaster Works 10/05/59
	With D5906		Test train (solo) to New England (12/05/59) with passenger stock returned to Doncaster light engine towing failed D5906.
		03/06/59	Test-run in multiple with D5908
			To Hornsey 08/06/59 (with D5900)
D5906	08/05/59	12/05/59	Noted at Doncaster Works 10/05/59
	With D5905		Test train (solo) to New England (12/05/59) with passenger stock, failed on arrival, hauled back to Doncaster by D5905 (light engines)
		09/06/59	To Hornsey 11/06/59
D5907	15/05/59	20/05/59	
D5908	29/05/59	03/06/59	Test-run in multiple with D5905.
D5909	19/06/59	23/06/59	
		14/07/59	Booked trials.

Notes:
1. BR CM&EE Doncaster 'Deliveries, Acceptance Trials, Despatch and Other Arrangements' information:

Plan dated 22/04/59 (for Period 27/04/59-01/05/59) (abridged):
D5904	27/04/59	In Works. Chromate and briquette treatment and trip gear.
	28/04/59	Path 2. 500tons.
	29/04/59	Path 3. 8 bogies. Peterborough.
	30/04/59	In Works. Exams.
	01/05/59	Despatch to Hornsey MPD.

Delivery and Acceptance • 53

Plan dated 22/04/59 (for Period 04/05/59-08/05/59) (abridged):

D5902	04/05/59	In Works. Chromate and briquette treatment. Make-off ATC wiring.
	05/05/59	Light engine trial to Harrogate and return. Available at Harrogate 4pm to 7pm.
	06/05/59	Path 3. 8 bogies. Peterborough.
	07/05/59	Path 2. 500tons.
	08/05/59	In Works. Exams & Trip Gear.

These plans clearly suggest more main-line trials were undertaken than reported by the BLS or CD&E6 in Section 8.

D5901, Doncaster Works (Paint Shop Yard), 23 May 1959. (RW [Transport Treasury])

Chapter 6
ALLOCATION HISTORY

6.1 Allocation History

D5900
Delivery from EE to Doncaster Works	22/05/59
Officially allocated to 34B Hornsey	22/05/59
Movement to 34B Hornsey	08/06/59
Re-allocated to 34G Finsbury Park	24/04/60
Out of service	25/01/63
Officially stored (u/s)	03/02/63 (4we 23/02/63)
Officially re-instated	28/07/63 (3we 03/08/63)
Transit to Vulcan Foundry for refurbishment	30/07/63
Officially stored (u/s)	01/12/63 (3we 21/12/63)
Re-instatement to 34G post-refurbishment	02/10/64
Withdrawn	30/12/68 (1we 04/01/69)

D5901
Delivery from EE to Doncaster Works	22/05/59
Officially allocated to 34B Hornsey	22/05/59
Movement to 34B Hornsey	xx/xx/59
Re-allocated to 34G Finsbury Park	24/04/60
Out of service	28/01/63
Officially stored (u/s)	03/02/63 (4we 23/02/63))
Officially re-instated	xx/xx/63 (4we23/03/63)
Transit to Vulcan Foundry for refurbishment	06/03/63
Officially stored (u/s)	01/12/63 (3we 21/12/63)
Re-instatement to 34G post-refurbishment	29/04/65
Re-allocated to Research Centre, Derby	xx/xx/xx (1we 060969) (on loan)
Withdrawn	07/12/69 (1we13/12/69) with effective permanent re-allocation to RCD

D5902
Delivery from EE to Doncaster Works	01/05/59
Officially allocated to 34B Hornsey	01/05/59
Movement to 34B Hornsey	12/05/59
Re-allocated to 34G Finsbury Park	24/04/60
Out of service	16/04/62
Officially stored (u/s)	25/11/62 (1we 01/12/62)
Officially re-instated	28/07/63 (3we 03/08/63)

Transit to Vulcan Foundry for refurbishment
Officially stored (u/s)   01/12/63 (3we 21/12/63)
Re-instatement to 34G post-refurbishment   27/11/64
Withdrawn   23/11/69 (1we 29/11/69)

D5903
Delivery from EE to Doncaster Works   17/04/59
Officially allocated to 34B Hornsey   17/04/59
Movement to 34B Hornsey   24/04/59
Re-allocated to 34G Finsbury Park   24/04/60
Out of service   14/02/62
Officially stored (u/s)   25/11/62 (1we 01/12/62)
Officially re-instated   28/07/63 (3we 03/08/63)
Transit to Vulcan Foundry for refurbishment   01/08/63
Officially stored (u/s)   01/12/63 (3we 21/12/63)
Re-instatement to 34G post-refurbishment   04/09/64
Withdrawn   30/12/68 (1we 04/01/69)

D5904
Delivery from EE to Doncaster Works   24/04/59
Officially allocated to 34B Hornsey   24/04/59
Movement to 34B Hornsey   01/05/59
Re-allocated to 34G Finsbury Park   24/04/60
Out of service   01/05/62
Officially stored (u/s)   25/11/62 (1we 01/12/62)
Officially re-instated   28/07/63 (3we 03/08/63)
Transit to Vulcan Foundry for refurbishment   01/08/63
Officially stored (u/s)   01/12/63 (3we 21/12/63)
Re-instatement to 34G post-refurbishment   01/07/64
Withdrawn   20/01/69 (1we 25/01/69)

D5905
Delivery from EE to Doncaster Works   08/05/59
Officially allocated to 34B Hornsey   08/05/59
Movement to 34B Hornsey   08/06/59
Re-allocated to 34G Finsbury Park   24/04/60
Out of service   13/06/63
Officially stored (u/s)   16/06/63
Officially re-instated   28/07/63 (3we 03/08/63)
Transit to Vulcan Foundry for refurbishment   xx/xx/63
Officially store (u/s)   01/12/63 (3we 21/12/63)
Re-instatement to 34G post-refurbishment   20/12/64
Withdrawn   14/02/71 (1we 20/02/71)

D5906
Delivery from EE to Doncaster Works   08/05/59
Officially allocated to 34B Hornsey   08/05/59
Movement to 34B Hornsey   11/06/59

Re-allocated to 34G Finsbury Park 24/04/60
Out of service 05/10/62
Officially stored (u/s) 25/11/62 (1we 01/12/62)
Officially re-instated 28/07/63 (3we 03/08/63)
Transit to Vulcan Foundry for refurbishment 27/07/63
Officially stored (u/s) 01/12/63 (3we21/12/63)
Re-instatement to 34G post-refurbishment 29/10/64
Withdrawn 30/09/68 (1we 05/10/68)

D5907
Delivery from EE to Doncaster Works 15/05/59
Officially allocated to 34B Hornsey 15/05/59
Movement to 34B Hornsey 22/05/59
Re-allocated to 34G Finsbury Park 24/04/60
Out of service 13/11/62
Officially stored (u/s) 25/11/62 (1we 01/12/62)
Officially re-instated 28/07/63 (3we 03/08/63)
Transit to Vulcan Foundry for refurbishment 30/07/63
Officially stored (u/s) 01/12/63 (3we 21/12/63)
Re-instatement to 34G post- refurbishment 31/03/65
Officially stored (u/s) 18/10/68
Withdrawn 20/10/68 (1we26/10/68)

D5908
Delivery from EE to Doncaster Works 29/05/59
Officially allocated to 34B Hornsey 29/05/59
Movement to 34B Hornsey 05/06/59
Re-allocated to 34G Finsbury Park 24/04/60
Out of service 23/08/62
Officially stored (u/s) 25/11/62 (1we 01/12/62)
Officially re-instated 28/07/63 (3we 03/08/63)
Transit to Vulcan Foundry for refurbishment 23/07/63
Officially stored (u/s) 01/12/63 (3we 21/12/63)
Re-instatement to 34G post-refurbishment 18/08/64
Withdrawn 09/03/69 (1we15/03/69)

D5909
Delivery from EE to Doncaster Works 19/06/59
Officially allocated to 34B Hornsey 19/06/59
Movement to 34B Hornsey xx/xx/59
Re-allocated to 34G Finsbury Park 24/04/60
Out of service 07/03/62
Officially stored (u/s) 25/11/62 (1we 01/12/62)
Officially re-instated 28/07/63 (1we 03/08/63)
Transit to Vulcan Foundry for refurbishment xx/xx/xx
Officially stored (u/s) 01/12/63 (3we 21/12/63)
Re-instatement to 34G post-refurbishment 16/07/64
Withdrawn 07/03/71 (1we13/03/71)

Notes:
1. English Electric delivery dates were always Fridays i.e. Vulcan Foundry's normal release day.
2. Blue items and red items (withdrawal dates) represent the only information which was released externally by the Rolling Stock Library. The D5901 transfer to Derby Research Centre only ever appeared in the Birmingham Railway Club *Circular*.
3. Brown items. Out of service dates as recorded in *Deltic Deadline* (DPS).
4. Purple items: Extracted from official availability returns circulated internally within BR Eastern Region (i.e. Availability by Classes of Locomotives and Power Cars" [BR.1712/444]).

"Availability by Classes of Locomotives and Power Cars" (BR.1712/444)
Average Weekday Position for selected weeks (Nos. of Locomotives):

Week-Ending (W/e)	S(u)	Net Op. Stock	Available	Not Available	
				Works	Depots
W/e 24/11/62	-	10	2	7	1
W/e 01/12/62	7	3	2	0	1
W/e 08/12/62	7	3	3	0	0
W/e 15/12/62	7	3	3	0	0
W/e 22/12/62	7	3	3	0	0
W/e 29/12/62	7	3	3	0	0
W/e 26/01/63	7	3	2	0	1
W/e 23/02/63	9	1	1	0	0
W/e 23/03/63	8	2	1	1	0
W/e 06/04/63	8	2	1	1	0
W/e 18/05/63	8	2	1	1	0
W/e 15/06/63	8	2	1	1	0
W/e 13/07/63	9	1	0	1	0
W/e 03/08/63	-	10	0	10	0
W/e 07/09/63	-	10	0	10	0
W/e 05/10/63	-	10	0	10	0
W/e 02/11/63	-	10	0	10	0
W/e 30/11/63	-	10	0	10	0
W/e 21/12/63	10	0	0	0	0

Abbreviations/explanations

S(u)	Stored unserviceable.
Net Op. Stock	Net Operating Stock
Not Available Works	In Works & Awaiting Works
Not Available Depots	Under/Awaiting Repairs or Exams

5. Green items: Transit to Vulcan Foundry (see Section 9.10).
6. Re-instatement after refurbishment. Dates from Diesel Locomotive Record Cards and Brian Webb (1982).

D5908 and D5907, 34B Hornsey, June 1959. (I.S. Carr [David Dunn Collection])

D5908 and D5907, 34B Hornsey, June 1959. Looking north. (I.S. Carr [David Dunn Collection])

6.2 Key Depots and Stabling Points

6.2.1 34B Hornsey

The western half of the old Hornsey steam depot (Roads 1-4) was partitioned off and designated for diesel usage and was kept as clean as practically possible. This was an interim measure pending construction of the purpose-built facility at Finsbury Park which opened in April 1960. Locomotives continued to be serviced at Hornsey until the site was turned over to diesel multiple unit (DMU) stabling in March 1969.

6.2.2 34D Hitchin

A custom-built two road dead-end diesel shed was opened in 1961 to service out-based Finsbury Park main-line locomotives and local shunters, together with a fuelling point and stabling facilities. Apart from major maintenance, Hitchin was effectively the operational depot for the 'Baby Deltics' throughout their lives.

Above left: D5907 and D5902, 34B Hornsey, 24 May 1959. Diesel half of the depot in the foreground and left, with the steam section to the right. (A.R. Valentine [David Dunn Collection])

Above right: D5905, 34B Hornsey, 29 June 1969. (R. Holland [David Dunn Collection])

Left: D5909, Hitchin s.p., Undated. Turntable. (Peter Ingarfill)

Below left: D5909, Hitchin s.p., 7 June 1970. (Kevin Hughes [RCTS Archive])

Below right: D5909, Hitchin s.p., 8 June 1969. (Peter Foster)

D5905, Hitchin s.p., July 1970. (Grahame Wareham)

6.2.3 34G Finsbury Park (Clarence Yard)

Finsbury Park was a custom-built diesel depot which was officially opened on 24 April 1960, although the May 1960 *Railway Observer* stated that 'Clarence Yard diesel depot was brought into use on 10 April'. It was a six-road dead-end shed, each road designed to accommodate three locomotives, with fuelling and stabling facilities. Maintenance and servicing work was undertaken at Finsbury Park with light servicing support provided by King's Cross, Hornsey and Hitchin.

In his book *The Trials and the Triumph* (2012), Tom Grieves was critical of the facilities available at Finsbury Park particular with regard to the maintenance of the production 'Deltic' and 'Baby Deltic' fleets:

A fundamental error in the clash of motive power depot requirements and chief mechanical engineering blindness, supported by the Board planning manager, denied the provision of an engine lift facility for the lightweight 'Deltic' engines of both Type 2 and Type 5. This cost the industry dearly in loss of availability by having to send locomotives to Doncaster (*or Stratford*) for engine change where the Work's five eight-hour day week conflicted with the Motive Power's 24-hour 7-day need to keep services running…

D5903, 34G Finsbury Park, 13 July 1966. (John Grey Turner)

On two occasions when availability was descending to a level that the depot threatened to reinstate steam on the outer suburban the following week, a spare engine was delivered by road from Doncaster and changed with the King's Cross breakdown crane, each occasion within an eight-hour shift. The Western with their diesel-hydraulics applied more common sense by providing engine lift facilities at both Old Oak and Laira, bringing Western diesel availability to near target level.

6.2.4 King's Cross s.p

This location included a single-track dead-end shed for minor servicing, together with fuelling and stabling facilities.

D5901, 34G Finsbury Park, 8 June 1969. Fuelling Point. (R. Holland [David Dunn Collection])

D5903, King's Cross s.p., 20 November 1968. (Author's Collection)

Chapter 7
MAJOR OVERHAULS AND REPAIRS

7.1 Stratford DRS

Right: **D5901 and D5909, Stratford DRS, 7 January 1962.** (Norman Browne [RCTS Archive])

Below left: **D5900, 30A Stratford, Undated.** Jubilee shed yard (south-west end). Possibly stored. (Colour-Rail)

Below right: **D5902, Stratford DRS (Works Yard), April 1970.** (Grahame Wareham)

7.2 Doncaster Works

Above left: **D5903 Doncaster Works (Erecting Shop). 9 January 1966.** (N.W. Skinner [David Dunn Collection])

Above right: **D5900 and D5506, Doncaster Works (Erecting Shop Yard), 20 April 1968.** D5900 in Works for an Unclassified repair (released 2 May) and D5506 present for conversion from Class 30 to Class 31 (released June 1968 after fitment of an English Electric engine). (Colour-Rail)

Left: **D5900, Doncaster Works (Erecting Shop), Undated.** (David Dunn Collection)

Doncaster Works Sightings (June 1964–December 1969).
N.B. Post-withdrawal sightings excluded.

Date	Note	Sightings
07/06/64		Nil
05/07/64		D5909 (Paint Shop Yard)
12/07/64		D5908/9 (both Paint Shop Yard)
27/07/64		D5908
08/08/64		Nil
22/08/64		D5903
03+04/10/64		Nil
11/10/64		Nil
01/11/64		D5902 (Paint Shop & area)
09/11/64	Cmt	D5900 (arrival date)
21/11/64		D5902
05/12/64		D5905 (Paint Shop)
06/12/64		D5905
13/12/64		D5905 (Paint Shop)
31/12/64		Nil.
10/01/65		Nil
23+24/01/65		Nil
07/02/65		Nil
27/02/65		Nil
07/03/65		Nil
13/03/65		D5907
14/03/65		D5907 (Paint Shop Yard)
03+04/04/65		Nil
24/04/65		D5901 (Paint Shop Yard)
02/05/65		D5903 (Works Yard)
09/05/65		D5903 (Erecting Shop Yard)
04/07/65		Nil
11+14/07/65		Nil
23/07/65	Cmt	D5900 (arrival date)
28/07/65	Cmt	D5900 (release date)
08/08/65		Nil
19/09/65		D5904
10/10/65		D5900 (Erecting Shop)
16/10/65		D5900
14/11/65		D5904
21/11/65		D5904 (Erecting Shop), D5908 (Works Yard)
01/12/65	Cmt	D5906 (arrival date)
05/12/65		D5904/8, plus D5906 (36A Doncaster)
10/12/65		D5908 (Erecting Shop)
09/01/66		D5903 (Erecting Shop)
06/02/66		Nil
12/02/66		D5905
19/02/66	Cmt	D5901 (arrival date)
03/04/66		D5902
17/04/66		D5908 (Erecting Shop)
01/05/66		D5908
05/06/66		Nil, plus D5904 (36A Doncaster)
07/05/66		Nil
21/05/66	Ph	D5905 (Erecting Shop Yard)
05/06/66		Nil
10/07/66		Nil
13/07/66	Cmt	D5901 (arrival date)
03/08/66		Nil
07/08/66		D5907
20/08/66		Nil
09/10/66		Nil
19/10/66	Cmt	D5900 (arrival date)
28/10/66	Cmt	D5900/7 (release date)
15/11/66	Cmt	D5909 (release date)
11/12/66		D5909 (rectification?)
15/12/66	Cmt	D5909 (release date)
30/12/66	Cmt	D5908 (arrival date)
08/01/67		D5900/8
05/02/67		D5904
05/03/67		D5903

Major Overhauls and Repairs • 65

Date	Note	Sightings
02/04/67		Nil
21/05/67		Nil
09/07/67		D5901/6
27/07/67		D5906/8
07/08/67		D5906 (Erecting Shop), D5908 (Works Yard)
23+24/09/67		D5906
01/10/67		D5906/8
15/10/67		D5901/8
03/12/67		D5901
04+07/01/68		D5908 (Works Yard), D5909 (Erecting Shop)
09/01/68		D5908/9 (both Erecting Shop)
11/01/68	DLRC	D5908 (Unclassified repair)
28/01/68		D5903/5. D5909 not listed.
04/02/68		D5903/5. D5909 not listed.
17/02/68		D5903/5/9
19/02/68	Ext	D5909 (Station Yard)
03/03/68		D5903/5 (both Works Yard)
20/03/68	DLRC	D5903 (Unclassified repair)
31/03/68		Nil
20/04/68		D5900 (Erecting Shop Yard)
02/05/68	DLRC	D5900 (Unclassified repair)
05/05/68		Nil
06+09/06/68		Nil
03/07/68	DLRC	D5907 (Unclassified repair)
07/07/68		Nil
04/08/68		Nil
23/08/68		D5900
19/09/68		Nil
21/09/68		Nil
09/11/68		Nil
26/11/68		D5909
01/12/68		D5909 (Works Yard)
10/12/68		D5905/9

Date	Note	Sightings
18/12/68	Cmt	D5905 (release date)
05/01/69		D5908
11/01/69		Nil
14/01/69	DLRC	D5908 (start of Unclassified repair)
18/01/69		D5908
21/01/69		D5908
23/02/69		D5908
09/03/69		Nil
10+20/04/69		Nil
15/06/69		Nil
13+26/07/69		Nil
23+24/08/69		Nil
27/09/69		Nil
02+09/11/69		Nil
13/12/69		Nil

Notes:
1. Abbreviations: Cmt: Magazine comment only, Ext: External Works sighting only; Ph: Photograph; DLRC: Diesel Locomotive Record Card entry.
   Each of the above are not, therefore, full visit reports.
2. Pre-rehabilitation, the vast majority of Works repairs were undertaken by Stratford DRS (Classified and Unclassified), with no sightings found of 'Baby Deltics' at Doncaster Works at all during this period (other than acceptance trials and D5905 in June 1963).
3. Post-rehabilitation: All Classified and most Unclassified work undertaken by Doncaster Works. The Unclassified work which was undertaken by Stratford DRS included radiator changes, traction motor changes, boiler attention, bogie cleaning, tyre-turning, etc., plus the cannibalisation of withdrawn locomotives to provide spares for remaining locomotives in service (inc D5901 in Departmental stock).
4. In an article entitled 'D5909 - How the Baby Became Blue' in 'Deltic Deadline' (*Journal of the Deltic Preservation Society*, Issue 29, October 1982) author Andy Wylie stated, 'With the normal shopping of the "Baby Deltics" suspended as from January 1968, no other member of the class was repainted and therefore D5909 remained the sole "blue Baby" to the end.'
   Work had started on D5909 in late December 1967/early January 1968 and it was outshopped on 16 February.

Chapter 8
LOCOMOTIVE HISTORIES

8.1 Explanation of Information Provided

8.1.1 Sightings: Primary and Secondary Sources

Locomotive histories have been built up for each of the 'Baby Deltic' locomotives, developed from primary sources (personal sightings and photographs) and secondary sources (magazine reports, archive depot/works visit sightings and other listings providing fully dated information). Due to space constraints, only the more significant sightings are reproduced here; however, information is provided to cover all Works sightings and most post-withdrawal observations.

Whilst Primary information is undoubtedly the best source of sighting details and associated anecdotal information, it is not without shortcomings; for example Stratford Works visits were frequently not explicit as regards to the exact location of locomotives (in Works or Works Yard) and locomotives in Works were often missed due to visits to Stratford being restricted to the depot only.

8.1.2 Works Information

Works information from Diesel Locomotive Record Cards has been found for six of the ten 'Baby Deltics' (D5900/3/4/6-8), but unfortunately these are incomplete with only limited shopping information recorded post-refurbishment from 1964/65. Fortunately, hand-written BR (ER) (King's Cross District) records fill in the pre-1964 gaps.

Other sources of Works information include:

- RCTS *Railway Observer* 'Stratford Works' reports provided monthly 'Ex-Works after heavy repair' information from October 1959 until October 1962 (Heavy Intermediate and Heavy Casual [power unit change] repairs). On comparison with the DLRC it is clear that the first two or three reports were incomplete, but they quickly settled down to provide comprehensive information. The 'late start' precluded any 'Baby Deltic' repair reports during the second half of 1959, but from the beginning of 1960s the RCTS reports both corroborated the DLRCs and the BR (King's Cross District) reports. The RCTS reports covered the period up to the point when 'Baby Deltic' heavy repairs at Stratford ceased in May 1962. Light Casual repairs were not recorded in the RCTS reports.
- Stratford DRS Records. Two separate and complementary sources of information cover locomotive throughput at Stratford DRS during the period November 1963 to March 1971 and highlight visits made to this establishment for such work as tyre-turning, bogie washing, etc.
- Sighting Reports. To achieve something close to a reasonably comprehensive history of Works visits by the 'Baby Deltics', particularly after refurbishment, required heavy reliance on Doncaster Works sighting information (see Section 7). There are undoubtedly still a number of visit omissions, but the sightings do contribute substantially to the story.

8.1.3 Information from B.R. 'Fires on Diesel Train Locomotives Reports' (1961-71)

All reported incidents are included (see Section 14.2).

8.1.4 Information from BR (ER) 'Statement of Diesel Casualties (Mechanical/Electrical)' King's Cross District Reports (December 1961-June 1963) (BR 33852/2)

Selected incidents are included (see also Section 9.7)

8.1.5 Other Secondary information

On certain occasions other secondary information from a

wide variety of sources has been included with the strict proviso that it was fully dated and/or provided additional commentary around already established primary information.

8.2 Data Presentation
Sighting Dates
The date format in the history logs is 'mmyy' or 'ddmmyy', as opposed to 'mm/yy' or 'dd/mm/yy', for clarity. The more conventional 'dd/mm/yy' format is, however, used in the locomotive 'Notes' (excepting quotes where the original published format is retained).

Text colour coding is used as follows:

Blue	Key dates (e.g. New, Works information, storage, withdrawal and disposal).

N.B. Disposal information is sourced from *Diesel & Electric Locomotives for Scrap* (A. Butlin) although this topic is subject to further debate in Sections 20.2 and 20.3.

Black	Locomotive sightings.
Purple	Dates when locomotives were NOT seen at a particular depot or works.
Red	Sightings information conflicts.

'Nil' refers to no sightings at the specified location.
'N/listed' means not listed on specified date.

Various abbreviations are used as follows:

Depot Abbreviations
30A: Stratford; **34B:** Hornsey; **34D:** Hitchin; **34G:** Finsbury Park.
Other Location Abbreviations:
KX: King's Cross; **PS:** Paint Shop (Doncaster Works); **SFDRS:** Stratford Diesel Repair Shop.

Works Information
References to Works visits in the following logs (e.g. DLRC: 260859-280859 N/C) specify 'Information Source', 'Date in Works' to 'Date out of Works' and 'Class of Repair' (if known) and/or 'Work done'.

Overhaul categories, or, Class of Repair:
Pre-1964: **HI** : Heavy Intermediate, **HC:** Heavy Casual, **LC:** Light Casual, **U, U/C or N/C**: Unclassified (unscheduled).

From 1964: **G**: General, **I**: Intermediate, **C:** Classified (repair level unspecified), **Rect**: Rectification (i.e. re-call to Works to correct faults), **U or U/C**: Unclassified (unscheduled).

Fires
Information from BR 'Fires on Diesel Train Locomotives Reports' are suffixed **FDTL**.

Casualties
Information from BR 'Statement of Diesel Casualties' are suffixed **BRCS.**

Other Abbreviations
Jct: Junction.
CSdgs: Carriage Sidings, **Sdgs:** Sidings.
LE: Light Engine, **Pass:** Passenger service.
N/B: Northbound, **S/B:** Southbound
STN: Special Traffic Notice,

Source Information
BRDoc: Handwritten BR record of Works Visits for King's Cross District locomotives.
Derived Rcds: Derived Works Records.
DLRC: Diesel Locomotive Record Cards.
SFDRSReps: Stratford DRS Reports kept by Arthur Nugent and Aubrey Rayment (September 1964 to February 1971).
BDP: Baby Deltic Project.
DD: "Deltic Deadline" (DPS).
And finally: Abbreviations of the sources used in the 'Notes' comments at the end of each Locomotive History are explained in the "Sources & References" section.

8.3 Locomotive Histories
Abridged histories are provided on the following pages.

D5900
EE/VF Works Nos.: 2377/D417

Above left: **D5900, 34B Hornsey, 1960.** (Grahame Wareham Collection)

Above right: **D5900, Location and date unknown.** (Peter Ingarfill)

Below left: **D5900, Wood Green, Undated.** GFY livery. Judging by the freshness of the full yellow front-end paintwork, this is probably around September 1968. Part of the Sherwood Green paintwork has also been re-applied. (Author's Collection)

Below right: **D5900 and D5903, 30A Stratford (alongside the 'New' Shed), 8 March 1969.** (Author's Collection)

New (date dispatched ex-EE): 220559.

Doncaster Works: 220559 (PS Yard)/260559 (main-line test run, in multiple with D5901)/ 310559.
DLRC: Doncaster: 220559-080659 Trials.

To 34B: 080659 (with D5905)

SFDRS: Nil. DLRC: 260859-280859 N/C (air-brake modifications). N/listed 300859.

SFDRS: Nil. DLRC: 170959-180959 HC. N/listed 130959.

34B: 200959
KX: 301259 (LE)

SFDRS: Nil. DLRC: 130160-200160 HC (Power-unit change).

34B: 310160
KX: 260460

SFDRS: Nil. DLRC: 170560-190560 HC (Power-unit change). N/listed 210560.

KX: 170660 (LE)/ 100860 (pass)/ 250860 (LE)

SFDRS: 180960. DLRC: 130960-071060 HC (Power-unit change).

SFDRS: 110261 (Erecting Shop)/180261. DLRC: 180161-040361 HI (No.2 Classified Repair) (75,380 miles recorded).

KX: 240661
Hitchin: 040861

SFDRS: 160961. DLRC: 160961-220961 HC (Power-unit change).

34D: 151061
Welwyn Garden City: 231261 (pass)
34G: 260162 (exam, cracked cylinder liner) (BRCS)

SFDRS: 250262 (Erecting Shop). DLRC: 300162-150362 HC (Power-unit change).

KX: 200362 (LE)
34D: 230462
34G: 290462
XX: 040562 (12.30 KX-Royston pass, failed [T/M flashover]) (BRCS)
34G: 060562

SFDRS: 130562. DLRC: 090562-120562 LC (Traction motor repairs).

XX: 270762 (assisted failed A3 60103 [13.00 ex-Harrogate-KX?] from Three Counties)
XX: 011162 (05.22 Hitchin-KX pass, failed [loss of oil, burst oil pipe]) (BRCS)
XX: 031162 (Finsbury Park-Hitchin LE, failed [lubricating oil pipe adrift]) (BRCS)
XX: 051162 (Hitchin-Hatfield freight, failed [lubricating oil pipe broken]) (BRCS)
XX: 051262 (06.00 Baldock-Royston e.c.s., failed [loss of coolant]) (BRCS)
XX: 140163 (Hitchin-Royston freight, failed [loss of coolant]) (BRCS)
34G: 200163
XX: 250163 (08.20 KX-Baldock pass, failed [cut out in traffic, suspected piston seizure]) (BRCS/DD)

Stored (u/s): 030263 (4we230263).
SFDRS/ 30A Stratford: 030263/ 030363/ 060363/ 160363/ 310363/ 070463/ 210463/ 280463/ 230663/ 070763

Re-instated: 3we030863 (280763).
Vulcan Foundry: Nil. DLRC: EE/VF: 300763-
Stored (u/s): 3we211263 (011263).
Vulcan Foundry: Nil. DLRC: EE/VF: 300763-xxxxxx.
Doncaster Works: 250964 (main-line test run).
DLRC: Doncaster Works: xxxxxx-021064 Trials.
Re-instated: 021064.

To 34G: 021064
KX: 021064 (arrived assisting D264 on 07.45 Sunderland-KX)
Cadwell (north of Hitchin): 031064 (S/B freight)
Huntingdon: 241064 (freight)
Cambridge: 271064 (freight for Whitemoor Yard)
34D: 310165

Hitchin South: 130465 (S/B freight)
Cambridge: 200565 (pass)
Stevenage: 040665 (08.00 KX-Baldock pass/ 09.51 Baldock-KX pass/ 17.15 KX-Royston pass/ 19.50 Royston-KX pass)
Chelmsford: 060665 (excursion ex-KX)

Doncaster Works: Arrived 230765 (towing D5603)/ Departed 280765 (towing D5064)

KX: 140865/ 080965
Hitchin-KX: 100965

Doncaster Works: 101065 (Erecting Shop)/ 161065

KX: 020966 (stabled)

Newark: 191066 (en route to Doncaster Works, with D9001)
Doncaster Works: Nil.
Grantham: 281066 (01.00 S/B with D5907)

34D: 111266

Doncaster Works: 080167

KX: 040267 (LE)
34D Hitchin: 090467/ 050867
Potters Bar: 130867 (engineer's train)

SFDRS: Nil. SFDRSReps: 101067-171067 (Radiators changed, Radiator fan gearbox to examine, both bogies out for exam for rough riding, bogie wash).

Doncaster Works: 200468 (Erecting Shop Yard).
DLRC: xxxxxx-020568 U/C.

XX: 130568 (ballast, with D5908)
Hertford North: 140568 (pass)

Stevenage: 290568 (19.20 Palace Gates-Whitemoor coal empties, with D5909)

SFDRS: Nil. SFDRSReps: 310568-010668 (Tyre-turning).

34G: 020668
Langley Jct: 130668 (Palace Gates-Whitemoor coal empties, with D5909)
34D: 200768
KX: 290768 (07.15 Royston-KX pass)

Doncaster Works: 230868. N/listed 040868 and 190968.

XX: 070968. GFY livery.
34D: 131068/ 171168
Wood Green: 131268 (07.15 Royston-KX pass, failed)

SFDRS: 191268/ 251268. SFDRSReps: 181268-241268 (Blower fan drive stripped for D5909).

Condemned: 1we040169 (301268).

30A: 190169/ 160269/ 080369 (alongside New Shed). N/listed 130369.
34G: 220369/ 230369
Ferme Park Down Sdgs: 090469
34B: 200469
34G: 310569/ 080669. N/listed 290669.

Transfer: 9Z13 Finsbury Park-Kettering: 260669 (with D5903/4)

Kettering Yard/Station Sdgs: 030769
G. Cohen, Kettering: 070769/ 100769 (whole)/ 160769 (whole)/ 200769 (whole)/ 070869 (part-cut)/ 140869 (engine block on mainframe, on bogies)/ 010969 (power unit only). N/listed 060769 and 300869.

Disposal: GCK 0669-0769 (D&ELfS).

Note:
1. Mileage at 01/10/62 (still in traffic) since new: 148,460; mileage at 06/01/67 since refurbishment: 107,970.

D5901
EE/VF Works Nos.: 2378/D418

Above left: **D5901, King's Cross, Undated.** (David Dunn Collection)

Above right: **D5901, King's Cross, 30 May 1962.** Servicing shed to the right of picture with a 'Peak' receiving attention. (Nigel Petre)

Below left: **D5901, Derby RTC, 4 October 1976.** '34G' still painted on the cab-side where the metal shed plates used to be positioned, over seven years after re-allocation to Derby Research. (Steve Thorpe)

Below right: **D5901, Doncaster Works (close to Scrapping Area), 14 November 1976.** (Adrian Booth)

New (date dispatched ex-EE): 220559.

Doncaster Works: 230559 (PS Yard)/ 240559/ 250559 (PS Yard)/ 260559 (main-line test run in multiple with D5900). Derived Record: Doncaster Works: 220559-xxxx59 Trials.

To 34B: xxxx59
34D: 040659
Holloway: 160659 (pass)
Stevenage: 060759 (15.57 Baldock-KX pass, failed at Knebworth, rescued by J6 64197)

SFDRS: Nil. BRDoc: 270759-300759 Unclassified (air-brake modifications).

KX: 030959/ 190959 (S/B pass)/ 250959 (LE)

SFDRS: 291159. BRDoc: 241159-031259 HC (Power-unit change). N/listed 101159.

KX: 271259

SFDRS: Nil. BRDoc: 160360-180360 HC (Power-unit change). N/listed 270260.

KX: 260460
Woolmer Green: 180560 (06.53 KX-Cambridge, failed at WG, propelled to Knebworth by 9F 92168 on freight)

SFDRS: 210560 (Works Yard)/ 220560/ 290560. BRDoc: 200560-010660 HC (Power-unit change). N/listed 050660.

KX: 210660 (S/B pass)/ 170960

SFDRS: 091060. BRDoc: 101060-221060 HC (Power-unit change).

34G: 121160

SFDRS: 180361 (Erecting Shop)/ 230361 (Erecting Shop). BRDoc: 240261-080461 HI (No.2 Classified Repair).

KX: 180461 (pass)
34G: 140561
Hitchin: 200561 (pass)

SFDRS: Nil. BRDoc: 280661-300661 HC (Power-unit change).

34D: 180661
Cambridge: 120761 (pass)
Hitchin: 190861 (pass)
Oakleigh Park: 210861 (S/B pass)
Hatfield: 141061 (pass)
34G: 231261

SFDRS: 070162/ 210162
Ex-Works after heavy repair: January 1962 (RO0362).

XX: 270362 (15.05 KX-Ely pass, failed [cylinder liner cracked]) (BRCS)

SFDRS: 080462 (Erecting Shop). BRDoc: 290362-260462 HC (Power-unit change).

34G: 290462
XX: 300462 (KX-Royston pass, failed [lubricating oil pipe to radiator pulled off]) (BRCS)
Letchworth: 010562 (06.35 Hitchin-Baldock pass, failed [lubricating oil pipe to radiator burst]) (BRCS)
KX: 300562 (LE)
Finsbury Park: 290862 (pass)
XX: 151062-071162 and 211162-041262 (Mondays to Fridays) (06.43 Baldock-KX pass)
XX: 220163 (06.38 Hitchin-Baldock pass, failed [lubricating oil pipe to radiator burst])) (BRCS)
XX: 220163 (21.57 New England-Hitchin freight, failed [lubricating oil pipe to radiator burst]) (BRCS)
XX: 280163 (Cracked crankcase joint) (DD)

30A Stratford: 030263

Stored (u/s): 030263 (4we230263).
Doncaster Works: 170263/ 030363 (Works Yard).
Departed 030363 (Stevenage Loco Soc 0363). N/listed 030263/ 100263 and 100363.

Vulcan Foundry: Arrived 060363 (BWebb)
Re-instated: 4we230363 (xxxxxx) (assumed to be D5901- see Section 9.7).
Vulcan Foundry: Nil.
Stored (u/s): 3we211263 (011263).
Vulcan Foundry: Nil. Derived Record: EE/VF: xxxxxx-090465. Departed Vulcan Foundry 090465 (with D6976).
Doncaster Works: 230465 (main-line test run)/ 240465 (Paint Shop Yard)/ 260465 (main-line test run).

Derived Record: Doncaster Works: 090465-290465 Trials.
Re-instated: 290465.

To 34G: 290465
GN Main-Line: 040565
Stevenage: 040665 (15.05 KX-Ely 'Buffet' pass/ 17.30 Ely-KX 'Buffet' pass/ 20.30 KX-Baldock pass)

SFDRS: Nil. SFDRSReps: 150665-170665 (No.1 bogie LH top spring changed).

KX-Hitchin: 070965
Broad Street: 080965
Royston: 100965

Newark: 190266 (en route to Doncaster Works, hauled by D9016)
Doncaster Works: Arrived 190266 (hauled in by D9016)

KX s.p.: 020366
Hitchin: 040766 (2B66)

Newark: 130766 (en route to Doncaster Works)
Doncaster Works: Arrived 130766 (under own power)

34D: 111266/ 090467/ 070567
KX: 250567 (N/B pass)

Doncaster Works: 090767

34D: 050867
Langley Jct: 071067 (engineer's train)

Doncaster Works: 151067/ 031267

Finsbury Park: 200368 (7B76 06.33 Ashburton Grove-Blackbridge rubbish train and 08.42 return (both with D5904), then 11.15 Ashburton Grove-Blackbridge (solo)
34G: 030468/ 060468

SFDRS: 280568

KX: 200668
Langley Jct: 230668 (engineer's train)
KX: 120768 (N/B pass)/ 180968
34D: 171168
Wood Green: 041268

Finsbury Park: 180169 (freight)
KX: 180169 (fuel for KX s.p.)

SFDRS: 140369/ 160469. SFDRSReps: 240169-170469 (Radiators changed [radiators ex-D5903 or D5904]).
N/listed 130369.

34D: 250469
Welwyn North: 040569 (S/B ballast, with D5902)
KX: 230569 (17.18 Moorgate-Hatfield) (all available Brush Type 2s used on e.c.s. workings for additional main-line services)
34G: 080669
Hadley Wood: 140869 (LE)
KX: 230869 (LE)

Transfer: 1we060969 (310869)-Research Centre, Derby (on loan).

Derby RTC/ Etches Park Holding Sidings: 120969/ 210969/ 280969/ 051069

Condemned: 1we 131269 (071269).

Derby RTC/ Etches Park: 110470/ 260470

Doncaster Works: 200570

Derby RTC/Etches Park: 300570/ 130670/ 250770
Toton: 020870
Derby RTC/Etches Park: 080870/ 130870/ 290870/ 070970/ 281070/ 250471/ 270771
Friargate line: 3we210871 (trials with Conflat wagon)
Derby RTC/Etches Park: 081171/ 051271
Derby Works: 111271
Derby RTC/Etches Park: 060272/ 120472/ 170472/ 210572/ 300572/ 040672/ 170672/ 010772/ 160772/ 230772/ 090872
Transfer: Derby RCD-Mickleover Test Track: 170872 (D8598 hauling S18521 and D5901)
Derby RTC/Etches Park: 010972/ 011072
Leicester: 051072
Wolverhampton High Level: 061072 (Tribometer)/ 161072 (Tribometer)/ 181072 (Tribometer)/ 201072 (Tribometer)
Derby RTC/Etches Park: 221072
Mill Hill Broadway: 231072
15A Leicester: 271072

Nottingham: 011272
Derby RTC/Etches Park: 071272 (stabled on Tribometer)/ 161272
Harpenden: 020173
Derby RTC/Etches Park: 020273 (Derby-Birmingham-Stafford-Crewe-Derby Tribometer)
Mill Hill Broadway: 130273 (test train)
Derby RTC/Etches Park: 180273
Coseley/Tamworth HL: 190273 (Tribometer)
Mill Hill Broadway: 200273 (test train)
Derby RTC/Etches Park: 230273 (stabled on Tribometer)/ 240273/ 180373
Crewe: 300373 (test train)/ 020473 (Derby-Birmingham NS-Stafford-Crewe-Derby test train)
Derby RTC/Etches Park: 080473/ 200473/ 200573/ 290573
Leicester: 030673 (Tribometer)
Derby RTC/Etches Park: 230673
Dore: 310773 (Tribometer)
Derby RTC/Etches Park: 050873/ 110873 (stabled on Tribometer)/ 130873/ 160873/ 250873/ 280873
Chesterfield: 300873
Derby RTC/Etches Park: 310873
Chesterfield: 040973 (Tribometer)
Clay Cross: 130973
Derby RTC/Etches Park: 290973
Dronfield: 041073 (Tribometer)
Derby RTC/Etches Park: 021173/ 041173
Stafford: 121173 (Tribometer)
Derby RTC/Etches Park: 081273
Crewe engine sidings: 070174
Stoke: 080174 (test train)
Derby RTC/Etches Park: 180174/ 260174/ 020274

Trent South: 230274 (LE)
Derby RTC/Etches Park: 100374
Stoke: 180374 ("Hermes" test coach)
Derby RTC/Etches Park: 270474/ 180574/ 270574/ 010674
Sheffield: 210674 (test train)
Derby RTC/Etches Park: 220674/ 300674/ 150774/ 230774/ 040874/ 310874/ 220974/ 280974/ 081074/ 131074/ 301174/ 311274/ 100175/ 250175/ 220375/ 290375/ 120475/ 270475/ 310575
Lincoln St. Marks: 180675 (Tribometer)/ 190675 (Tribometer)/ 200675 (Tribometer)/ 290675 (Test Train)
Derby RTC/Etches Park: 160875
Bedford Jct: 190875 (Tribometer)
Derby RTC/Etches Park: 300875/ 070975/ 111275

Transfer Derby-Doncaster: 180276 (to Tinsley - see below)

41A Tinsley: 240276
36A Doncaster: 290276/ 070376

Doncaster Works: 280376/ 040476 (Works Yard)/ 100476/ 250476/ 020576/ 160576/ 230576/ 060676/ 130676/ 210676 (Dismantling Shop Yard; whole)/ 110776 (Works Yard)/ 250776/ 010876 (Scrap Line)/ 150876/ 220876/ 120976/ 031076 (Works Yard)/ 171076/ 311076/ 071176 (Works Yard)/ 211176/ 281176/ 051276 (Works Yard)/ 121276/ 090177 (Works Yard)/ 160177/ 230177/ 270177.
N/listed 060277.

Disposal: Doncaster Works 0277 (D&ELfS).

Note:
1. Mileage at 01/10/62 (still in traffic) since new: 146,620; mileage at 06/01/67 since refurbishment: 88,160.

D5902
EE/VF Works Nos.: 2379/D419

Above left: D5902, King's Cross, 23 April 1960. Radiator grille cover fitted. (Rail-Online)

Above right: D5902, Stratford DRS, 1961. (Grahame Wareham Collection)

Below left: D5902, Wood Green, 1961. (Geoff Sharpe [(Grahame Wareham Collection)])

Below right: D5902, 30A Stratford, Undated (1970). 'Modified' front end following minor accident damage. Withdrawn; source of spares for remaining locomotives in operational fleet. (W. Prescott [Colour-Rail])

Chester: 010459 (test run from Newton-le-Willows to Chester & return with one coach)
Garstang: 030459 (9-coach test run from Newton-le-Willows to Penrith & return)

New (date dispatched ex-EE): 010559.

Doncaster Works: 030559 (PS Yard)/ 060559 (main-line test run)/ 100559. Derived Record: Doncaster Works: 010559-120559 Trials.

To 34B Hornsey: 120559 (with D6103)
34B: 240559
34D: 040659

SFDRS: 070659 (Works Yard)

Woolmer Green: 080759 (N/B pass)

SFDRS: 160859 (Works Yard). BRDoc: 140859-150859 UC (air-brake modifications). N/listed 300859.

KX: 090959 (LE)/ 120959 (LE)
34B: 200959

SFDRS: Nil. BRDoc: 300959-011059 HC (Power-unit change?). N/listed 260959.

SFDRS: Nil. BRDoc: 141059-16 or 181059 HC (Power-unit change).

SFDRS: 010160 (Diesel Shop)/ 030160 (Erecting Shop). BRDoc: 221259-070160 HC (Power-unit change).

SFDRS: Nil. BRDoc: 220360-250360 HC (Power-unit change). N/listed 270260.

KX: 220460 (pass)/ 230460 (LE)/ 260460

SFDRS: 210560 (Erecting Shop)/ 220560. BRDoc: 170560-270560 HC (Power-unit change).

SFDRS: 110960. BRDoc: 040860-140960 HC (Power-unit change).

SFDRS: 051160. BRDoc: 290960-121160 HC (Power-unit change).

SFDRS: 110261 (Erecting Shop)/ 180261. BRDoc: 090261-250361 HI (No.2 Classified Repair).

KX: 080361/ 150361
Ashwell: 230561 (assisted D5903 on 17.33 Ely-KX pass from Ashwell to KX)
34D: 180661
Hitchin: 310761 (pass)
KX: 071061
34D: 151061
34G: 231261

SFDRS: 210162. BRDoc: 291261-240262 HC (Power-unit change).

XX: 060362 (07.05 Royston-KX pass, failed [piston seizure]) (BRCS)

SFDRS: Nil. BRDoc: 070362-060462 HC (Power-unit change).

34G: 080462
XX: 160462 (13.30 KX-Baldock pass, failed [cracked cylinder liners and con rod damage]) (BRCS/DD)

34G: 290462
SFDRS/ 30A Stratford: 290462/ 130562/ 270562/ 260762/ 280762/ 290762/ 120862 (Works Yard)/ 170862/ 260862 (Works Yard)/ 080962/ 061062/ 071062
Stored (u/s): we011262 (251162).
SFDRS/ 30A Stratford: 081262 (Works Yard)/ 030263/ 030363/ 060363/ 160363/ 310363/ 070463/ 210463/ 280463/ 230663/ 070763/ 200763
Cambridge: 200763 (hauled Stratford-Cambridge by D6711, with D5908)

Reinstated: 3we030863 (280763).
Vulcan Foundry: Nil.
Store (u/s): 3we211263 (011263).
Vulcan Foundry: Nil.
Doncaster Works: 011164 (Paint Shop & area)/ 041164 (main-line test run)/ 201164 (main-line test run)/ 211164/ 241164 (main-line test run). Derived Record: Doncaster Works: xxxxxx-271164 Trials.
Re-instated: 271164.

To 34G: 271164
GN Main-Line: 020465 (up freight, with D5643)
Stevenage: 040665 (Class 8 freight, then LE to Hitchin)
34D: 030765
Hitchin: 290765 (assisting D9013 on down 'Flying Scotsman', from Hitchin
Royston: 070965

Doncaster Works: 030466

KX: 120766 (LE)

SFDRS: Nil. SFDRSReps: 011266-071266 (Radiators changed).

34G: 111266
34D: 090467
Stevenage Old Town: 110467 (Hitchin-Ferme Park freight, derailed at Stevenage Up Yard; re-railed and returned to 34D Hitchin LE)
34D: 050867
Potters Bar: 130867 (engineer's train)
GN Main-Line: 090168 (down Esso tanks, with D5903)
KX s.p.: 090368
Welwyn Garden City: 110368 (06.41 Finsbury Park-WGC pass; exhaust fire, loco removed at WGC, local fire brigade in attendence)
Broad Street: 290368 (pass)
34G: 030468/ 050568
34B: 020668
KX: 010768
34D: 200768
34G: 060868

SFDRS: Nil. SFDRSReps: 130868-150868 (Tyre-turning).

Hitchin: 200868 (engineer's train)
34D: 131068/ 171168/ 211268/ 180169

SFDRS: Nil. SFDRSReps: 110269-140269 (Main steam pipe under No.1 bogie leaking).

34G: 230269/ 220369
KX: 020469 (N/B pass)
Welwyn North: 040569 (S/B ballast, with D5901)
KX: 210569 (S/B e.c.s.)/ 220569
Trumpington: 060669 (10.15 Hitchin-Chesterton Jct ballast, failed at Trumpington, hauled to Cambridge)
Cambridge: 060669 (failed, hauled by D5909 from Cambridge to Hitchin LE)

SFDRS: Nil. SFDRSReps: 200669-270669 (Auxiliary drive shaft and blower motor changed [both stripped from D5908]).

34G: 290669
Welwyn Garden City: 010769 (pass)
KX: 180869 (08.43 KX-Baldock pass)
34D: 180869/ 221169
Moorgate: 290869 (17.18 Moorgate-Hatfield; "the first 'Baby Deltic' to be noted on the Widened Lines service for many years.")
Peterborough: 060969 (down freight, returned south LE with Class 31)
34B: 111069
Chesterton Jct: 141069
GN Main-Line: 041169 (pass)
34D: 161169 (waiting for works attention due to loose tyres)/ 221169

Condemned: 1we291169 (231169).

34G: 200170/ 260170. Remained at 34G until end-0170 (David Percival). N/listed 301169/ 261269.

SFDRS: 070270(Erecting Shop)/ 200270/ 150370/ 190370 (Works Yard). N/listed 080270 (presumably in Erecting Shop). SFDRSReps: 260170-270270 (Traction motors removed and sent to Doncaster Works) and 120370-190370 (Radiators stripped for D5905).

30A Stratford: 290370/ 050470/ 200470/ 240470 (alongside New Shed)/ 260470 (alongside New Shed)/ 020570/ 230570/ 030670 (by New Shed)/ 070670 (alongside New Shed)/ 050770/ 010870 (alongside New Shed)/ 080870/ 120870/ 190870 (labelled for Cohen's, Kettering - see note 2)/ 220870 (by New Shed)/ 250870/ 020970 (despatched to Cohen, Kettering)

Transfer Stratford-Kettering: 020970

Kettering Yard: 030970

G. Cohen, Kettering: 230970/ 241070/ 121170 (whole)/ 141170. N/listed 200471.

Disposal: GCK 0970 (D&ELfS).

Notes:
1. Mileage at 01/10/62 (stored by this date) since new: 109,040; mileage at 07/10/66 and 06/01/67 (since refurbishment): 91,860 and 102,950 respectively.
2. '5902 (ex-34G), which has for some time been stored alongside the New Shed, was noted on 19th September labelled to Cohen's of Kettering.' (RO1170). N.B. Subsequent notes in paragraph relate to August 1970.

D5903
EE/VF Works Nos.: 2380/D420

Above left: D5903, 34B Hornsey, May 1959. Radiator grille cover fitted. (P.H. Groom)

Above right: D5903 (with D5902), Stratford DRS (Works Yard), 17 August 1962. Radiator grille cover removed. Both effectively, but not officially, stored unserviceable; neither would work again until 1964. (David Dippie [David Dunn Collection])

Below left: D5903, Doncaster Works (Erecting Shop Yard), 9 May 1965. (R. Mabey [Colour-Rail])

Below right: D5903, 34G Finsbury Park, circa 1968. GFY livery. (Rail-Online)

Transfer: Vulcan Foundry to Doncaster Works: 070459 (for weighing)

Doncaster Works: 080459 (weighing tests)

Transfer: Doncaster Works-Vulcan Foundry: 090459

New (date dispatched ex-EE): 170459.

Doncaster Works: 190459/ 210459 (main-line test run)/ 240459 (PS Yard). DLRC: 170459-240459 Trials.

To 34B: 240459
Royston: 050559 (Royston-KX pass)
34B: 080559
34D: 230559/ 040659
New Barnet: 190759 (pass)

SFDRS: Nil. DLRC: 240759-280759 NC (air-brake modifications).

Wymondley (nr. Hitchin): 170859 (N/B pass)
Hatfield: 040959 (Cambridge-KX pass)

SFDRS: Nil. DLRC: 041159-071159 HC (Power-unit change). N/listed 101159.

SFDRS: Nil. DLRC: 021259-051259 HC (Power-unit change).

Whitemoor Yard: 120260 (Whitemoor-Bury St. Edmonds coal train tests, with D5905)

SFDRS: 200260 (Works Yard). DLRC: 180260-190260 HC (Power-unit change). N/listed 270260.

KX: 260460
Greenwood: 310560 (18.39 Baldock-KX pass)

SFDRS: 110960/ 180960. DLRC: 040860 (also listed as 220860)-240960 HC (Power-unit change).

34G: 011060

SFDRS: 080161/ 110261 (Erecting Shop)/ 150161. DLRC: 050161-170261 HI (No.2 Classified Repair) (70,750 recorded miles).

SFDRS: Nil. DLRC: 080361-080361 NC (tyre-turning).

SFDRS: 180361 (Works Yard)/ 230361 (Erecting Shop Yard). DLRC: 150361-300361 NC (turbine bearings).

34G: 070561
Ashwell: 230561 (17.33 Ely-KX pass, failed at Ashwell, assisted by D5902 to KX)
34D: 180661
KX: 290661 (pass)
Hitchin: 040861
Sandy: 280861 (18.00 Huntingdon-Hitchin pass)
KX: 071061
New Southgate: 231061 (pass)

SFDRS: Nil. DLRC: 011161-041161 HC (Power-unit change).

SFDRS: 070162/ 210162. DLRC: 061261-070262 HC (Power-unit Change).

XX: 130262 (17.40 KX-Baldock pass, failed [loss of coolant water, piston seized]) (BRCS)

SFDRS/ 30A Stratford: 250262 (Works Yard)/ 080462 (Works Yard)/ 150462 (Erecting Shop)/ 290462/ 270562/ 120862 (Works Yard)/ 170862/ 260862 (Works Yard)/ 080962/ 061062/ 071062/ 211062. DLRC: 160262-261162 Waiting engine. N/listed 260762/ 280762.
Stored (u/s): 1we011262 (251162).
SFDRS/ 30A Stratford: 081262 (Works Yard)/ 030263/ 030363/ 060363/ 160363/ 310363/ 070463/ 210463/ 280463/ 230663/ 070763

Re-instated: 3we030863 (280763).
Vulcan Foundry: Nil. DLRC: 010863-
Stored (u/s): 3we211263 (011263).
Vulcan Foundry: Nil. DLRC: EE/VF: 010863-xxxxxx *(200864).*
Manchester Victoria: 200864 (LE in transit EE/VF to Doncaster Works)
Doncaster Works: 220864/ 240864 (main-line test run)/ 260864 (main-line test run). DLRC: Doncaster Works: xxxxxx *(200864)*-040964 Trials.
Re-instated: 040964.

To 34G: 040964
Hitchin: 100964

Cambridge: 190964 (pass)
Hitchin: 091064 (replaced D1568 on down "Aberdonian" at Hitchin, taking train to Peterborough)

SFDRS: 151164. SFDRSReps: 161164-201164 (radiators changed).

34D: 310165

Doncaster Works: 020565 (Works Yard)/ 090565 (Works Yard)

Stevenage: 040665 (07.45 Cambridge- KX pass)
34D: 030765
KX s.p.: 310765
KX: 070965

Doncaster Works: 090166 (Erecting Shop). N/listed 051265 and 090266.

Knebworth: 230666 (N/B pass)
34G: 130766
XX: 280766 (fire damage; not severe) (FDTL)
34D: 070866
Cambridge: 290866 ('Cambridge Buffet Express' to KX, with D5908)/ 300866 (ditto)/ 310866 (ditto)

SFDRS: Nil. SFDRSReps: 121066-181066 (radiators changed).

34G: 231066
34D: 111266

Doncaster Works: 050367. N/listed 050267 and 020467.

XX: 140367. GFY livery.
34D: 090467/ 250567/ 050867
Potters Bar: 130867 (engineer's train)
Hadley Wood: 081067 (ballast, with D5905)
34G: 171267
GN Main-Line: 090168 (down Esso tank train, with D5902)

Doncaster Works: 280168/ 040268/ 170268/ 030368 (Works Yard). DLRC: Doncaster Works: xxxxxx-200368 UC.

34G: 230368
Langley Jct: 300568 (Palace Gates-Whitemoor coal empties, with D5615)
Brookman's Park: 060668 (LE)
34G: 200768

SFDRS: Nil. SFDRSReps: 010868-030868 (bogie wash).

KX: 150868 (07.15 Royston-KX pass)
Peterborough: 071068 (LE)
34D: 171168
KX s.p.: 201168
Wood Green: 041268
34G Finsbury Park: 071268/ 081268

Condemned: 1we040169 (301268).

34G: 180169. N/listed 020369.

SFDRS: 080369 (alongside New Shed). SFDRSReps: 030269-xxxxxx (Radiators stripped for D5901; entry crossed out so work may have been cancelled, or, may reflect that loco was already withdrawn). N/listed 130369.

34G: 220369
Ferme Park Down Sdgs: 090469
34B: 200469
34G: 310569/ 080669

Transfer: 9Z13 Finsbury Park-Kettering: 260669 (with D5900/4)

Kettering Yard/Station Sidings: 030769/ 200769/ 210769 (awaiting entry to Cohen's Yard)
G. Cohen, Kettering: 070869 (whole)/ 140869 (whole)/ 300869 (whole)/ 010969/ 070969/ 100969/ 210969 (whole)/ 121069. N/listed 241069/ 261069.

Disposal: GCK 0669-1069 (D&ELfS).

Note:
1. Mileage at 01/10/62 (stored by this date) since new: 113,650; mileage at 06/01/67 since refurbishment: 116,340.

D5904

EE/VF Works Nos.: 2381/D421

Above left: D5904, 34B Hornsey, 19 March 1960. Radiator grille cover fitted. (N. Browne [Colour-Rail])

Above right: D5904, King's Cross, 18 June 1965. (Bill Reed [David Dunn Collection])

Below left: D5904, King's Cross, Undated. (Bill Reed [David Dunn Collection])

Below right: D5904, King's Cross, 29 July 1968. GFY livery. (David Percival)

New (date dispatched ex-EE): 240459.

Vulcan Bank: 230459 (LE)
Doncaster Works: 240459 / 290459 (main-line test run).
DLRC: 240469-010559 Trials.

To 34B: 010559
Broad Street: 120559 (16.38 Broad Street-Baldock pass)
34D: 040659
Holloway: 160659 (e.c.s.)

SFDRS: Nil. DLRC: 080759-090759 HC (Power-unit change by EE staff).

Hornsey: 130759 (S/B pass)

SFDRS: Nil. DLRC: 170859-190859 NC (air-brake modifications).

Cambridge: 260859
KX: 040959 (N/B pass)

SFDRS: Nil. DLRC: 151259-221259 HC (Power-unit change).

34B: 271259
KX: 291259
34B: 190360

SFDRS: 130460 (Diesel Shop). DLRC: 120460-140460 HC (Power-unit change).

Hitchin-KX: 260460
34D: 070660

SFDRS: 260660. DLRC: 090660-250660 HC (Power-unit change).

SFDRS: 110960 / 180960 / 091060. DLRC: 160860-151060 HC (Power-unit change).

34G: 121160

SFDRS: 180361 (Erecting Shop) / 230361 (Erecting Shop). DLRC: 100361-200461 HI (No.2 Classified Repair) (75,700 recorded miles).

XX: 080661 (failed, fractured cylinder liner)

SFDRS: Nil. DLRC: 090661-170661 HC (Power-unit change).

Hitchin: 040861
Potters Bar: 020961 (KX-Cambridge pass) / 110961 (pass)
34G: 011061
KX: 071061
34D: 151061
Hadley Wood: 191061 (pass)

SFDRS: Nil. DLRC: 011161-041161 HC (Power-unit change).

XX: 071261 (20.30 KX-Baldock pass, failed [lubricating oil pipe burst]) (BRCS)
XX: 030162 (06.32 Baldock-KX pass, failed at Hitchin [lubricating oil pipe burst]) (BRCS)

SFDRS: 250262 (Works Yard). DLRC: 140262-310362 HC (Power-unit change).

XX: 100462 (06.00 Baldock-Royston e.c.s, failed [lubricatind oil pipe burst]) (BRCS)
34G: 150462
New Barnet: 190462 (engineer's train)
34D: 230462
Welwyn: 260462 (pass)
Welwyn North Tunnel: 280462 (pass)
Welwyn Tunnel: 010562 (07.05 Royston-KX pass, failed [broken con rods and fractured crankcase], rescued by V2 60906) (BRCS)

SFDRS/30A Stratford: 270562 / 260762 / 280762 / 290762 / 120862 (Works Yard) / 260862 (Works Yard) / 080962 / 061062 / 071062 / 211062 / 041162. DLRC: 020562-261162 (waiting engine).
Stored (u/s): 1we011262 (251162).
SFDRS/30A Stratford: 081262 (Works Yard) / 030263 / 030363 / 060363 / 160363 / 310363 / 070463 / 210463 / 280463 / 230663 / 070763

Re-instated: 3we030863 (280763).
Vulcan Foundry: Nil. DLRC: 010863-
Stored (u/s): 3we211263 (011263).
Vulcan Foundry: Nil. DLRC: 010863-220664.
Doncaster Works: 250664 (main-line test run). DLRC: 220664-010764 Trials.
Reinstated: 010764.

To 34G: 010764
Arlesey: 010764 (LE)
KX: 020764 (05.44 Baldock-KX pass/ 08.20 KX-Baldock pass, then freight trips from Hitchin to Letchworth and Cambridge)/ 030764 (08.20 KX-Baldock pass)
Knebworth: 040764 (15.05 KX-Cambridge/Ely 'Buffet'/ 17.43 Ely/Cambridge-KX)
KX: 060764
Knebworth: 140764 (08.20 KX-Baldock pass)
KX: 070864 (LE)
KX-Hitchin: 100964

SFDRS: Nil. SFDRSReps (Arthur Nugent article): 211064-xxxxxx (radiators changed).

34D: 310165

SFDRS: Nil. SFDRSReps: 200365-250365 (radiators changed).

KX: 150565
Stevenage: 040665 (09.04 KX-Cambridge 'Buffet' pass/ 11.00 Cambridge-KX 'Buffet' pass/ 14.30 KX-Royston pass/ 17.13 Cambridge-KX pass)
KX: 060665 (excursion ex-KX)/ 180665 (N/B pass)
34D: 030765 (Steam Shed)

Doncaster Works: 190965. N/listed 161065.

Doncaster Works: 141165/ 211165 (Erecting Shop)/ 051265. N/listed 161065.

KX: 080166/ 300466

36A Doncaster: 050666

Hitchin: 180766 (LE)
34D: 111266
KX: 120167 (LE)

Doncaster Works: 050267
Newark: 110267 (en route to 34G from Doncaster Works with D9021, GFY livery)

KX s.p.: 140267. GFY livery.

34D: 090467
XX: 260467 (fire damage; severity not known) (FDTL)
34D: 050867
XX: 071167 (fire damage; not severe) (FDTL)
KX: 031267 (e.c.s.)
Baldock/Royston: 120268 (exhaust fire, attended to by local fire brigade)
KX: 120368
Finsbury Park: 200368 (7B76 06.33 Ashburton Grove-Blackbridge rubbish train and 08.42 return (both with D5901)
Cuffley: 010468 (17.15 Broad Street-Hertford passenger; exhaust fire, extinguished by driver, loco continued journey)
34G: 060468
Langley Jct: 230668 (engineer's train)
34D: 200768
KX: 290768 (LE)/ 150868 (07.48 Hertford-KX pass)
34D: 131068
34G: 271068
34D: 171168/ 071268
34G: 180169 (stored)

Condemned: 1we250169 (200169).

34G: 020269

SFDRS: Nil. SFDRS (Arthur Nugent article): 150269-xxxxxx (radiators stripped for D5901).

34G: 020369/ 220369
Ferme Park Down Sdgs: 090469
34B: 200469
34G: 310569/ 080669. N/listed 290669.

Transfer: 9Z13 Finsbury Park-Kettering: 260669 (with D5900/3)

Kettering Yard/Station Sdgs: 030769
G. Cohen, Kettering: 070769/ 100769 (whole)/ 160769 (whole)/ 200769 (whole). N/listed 060769 and 070869/ 010969.

Disposal: GCK 0669-0769 (D&ELfS).

Note:
1. Mileage at 01/10/62 (stored by this date) since new: 128,360; mileage at 06/01/67 since refurbishment: 124,070.

D5905
EE/VF Works Nos.: 2382/D422

Above left: D5905 Stratford DRS, 1960. (Grahame Wareham Collection)

Above right: D5905, Doncaster Works, 6 June 1963. (A. Ferguson [Colour-Rail])

Below left: D5905, 34G Finsbury Park, 18 July 1965. (Ray Simpson)

Below right: D5905, Hitchin s.p., 26 September 1970. (Courtney Haydon [RCTS Archive])

New (date dispatched ex-EE): 080559 (with D5906).

Doncaster Works: 100559/ 120559 (main-line test run to New England with passenger stock, returned light-engine to Doncaster hauling failed D5906)/ 030659 (main-line test run). Derived Record: Doncaster Works: 080559-080659 Trials.

To 34B: 080659 (with D5900)
KX: 290759 (LE)

SFDRS: Nil. BRDoc: 100859-130859 NC(air-brake modifications).

Southend Victoria: xx0859 (excursion)

SFDRS: 291159. BRDoc: 241159-111259 HC (Power-unit change).

SFDRS: Nil. BRDoc: 150160-260160 HC (Power-unit change).

Whitemoor Yard: 120260 (Whitemoor-Bury St. Edmonds coal train tests, with D5903)
Cambridge: 230460 (pass)
KX: 260460

SFDRS: 260660 (Works Yard). BRDoc: 230660-290760 HC (Power-unit change). N/listed 050660/ 070760.

KX: 170960

SFDRS: 051160. BRDoc: 201060-191160 HC (Power-unit change).

Biggleswade: 040261 (15.30 (SO) Biggleswade-Hitchin)
34G: 120261
KX: 250261

SFDRS: 300461 (Erecting Shop Yard). BRDoc: 250361-290461 HI (No.2 Classified Repair).

KX: 130561

SFDRS: Nil. BRDoc: 310561-100661 HC (Power-unit change).

34D: 180661

SFDRS: Nil. BRDoc: 140761-200761 HC (Power-unit change).

Hitchin: 040861
Potters Bar: 110961 (freight)
Hitchin: 120961 (pass)
34D: 151061
XX: 120362 (19.47 Royston-KX pass, failed [cracked cylinder liner]) (BRCS)

SFDRS: 080462 (Works Yard)/ 150462 (Erecting Shop)/ 290462. BRDoc: 140362-040562 HC (Power-unit change).

34G: 060562
XX: 170762 (Hitchin-Cambridge freight, failed [flames from exhaust]) (BRCS)
XX: 170862 (KX e.c.s. duties, failed [traction motor blower adrift]) (BRCS)
Potters Bar: 281262 (13.05 KX-Cambridge)
34G: 200163/ 310363
XX: 050463 (09.30 KX-Baldock pass, failed [flashover on Nos. 1 & 3 T/M]) (BRCS)
KX: 030563 (down 'Cambridge Buffet Express')
Langley Jct: 040563 (15.05 KX-Cambridge pass)
KX: 240563 (18.39 KX-Baldock pass, later in day in collision with Brush Type 2 and taken out of traffic)

Doncaster Works: 060663/ xxxx63. BRDoc: 290563-070663 LC (collision damage repairs). N/listed 090663.

KX: 100663 (18.39 KX-Baldock pass)
KX/New Southgate: 110663 (18.39 KX-Royston pass)
XX: 130663 (Metal found in 'chip trap') (DD)
34G: 140663 (on 'failed loco siding', north end of Clarence Yard)

Transfer to EE/VF: 150663 (DD).
Retford: 150663 (being taken north by D5069)

Doncaster Works: Nil. N.B. D5069 present at Doncaster Works on 070763.

Stored (s/u): 160663.

30A Stratford: 070763

Re-instated: 3we030863 (280763).
Vulcan Foundry: Nil.

86 • ENGLISH ELECTRIC TYPE 2 BO-BO 'BABY DELTIC' LOCOMOTIVES

Stored (u/s): 3we211263 (01/12/63).
Vulcan Foundry: Nil. Engine No.388 installed (BDP).
Doncaster Works: Arrived 291164/ 051264 (Paint Shop)/ 061264/ 091264 (main-line test run)/ 111264 (main-line test run)/ 131264 (Paint Shop). Derived Record: Doncaster Works: xxxxxx-201264 Trials.
Re-instated: 201264.

To 34G: 201264
KX: 010165
XX: 220265 (failed; loss of lub oil)
Stevenage: 040665 (09.35 KX-Baldock pass/ 16.36 Hitchin-KX pass/ 18.39 KX-Royston pass)
34D: 030765
34G: 110765
Hitchin: 070965
Finsbury Park: 080965
KX: 100965
34B: 160166

Doncaster Works: 120266

34B: 200266
KX: 300466

Doncaster Works: 210566 (Erecting Shop Yard). N/listed 070566. Installed engine No.388 reached 4000hrs (contracted overhaul life) on 050566; engine removed at Doncaster for repairs by Napier and subsequently re-installed in D5909 (BDP).

Finsbury Park: 030666 (pass)

SFDRS: Nil. SFDRSReps: 250766-040866 (No.1 Traction Motor).

Wood Green: 230866 (pass)
KX-Potters Bar: 030966
XX: 101066 (fire damage; not severe) (FDTL)
34D: 111266
Stevenage: 110267 (11.55 Cambridge-KX pass)
KX: 180267 (pass)
34D: 050867
Potters Bar: 130867 (engineer's train)
Hadley Wood: 081067 (ballast, with D5903)

Doncaster Works: 280168/ 040268/ 270268/ 030368 (Works Yard)

KX: 230468 (N/B pass)

SFDRS: Nil. SFDRSReps: 070568-090568 (bogie wash).

Hertford North: 030768 (pass)
34B: 200768
Woolmer Green: 170868
Wood Green: 220868 (N/B pass ex-Broad Street)
Sandy: 270868 (freight)
34B: 171168

SFDRS: Nil. SFDRSReps: 191168-201168 (tyre-turning).

Hadley Wood: 031268 (07.15 Roysyon-KX pass, caught fire)

Doncaster Works: 101268. Engine No. 388 installed following repairs/overhaul by Napier; released into traffic on 181268 (BDP).

34D: 180169
Harringay West: 170269 (pass)
GN Main-Line: 200269 ("working residentials")
34D: 230269
KX: 180369 S/B pass)/ 280369 (S/B pass)/ 170469 (S/B pass)
Knebworth: 180469 (07.58 Royston-KX)
34G: 200469
34D: 250469
34B: 290669
KX: 150869 (17.39 KX-Peterborough pass, in lieu of Class 47, with D5909)
Newark: 020969 (arrived LE, then worked pigeon special)
GN Main-Line: 241069 (freight)
Welwyn Garden City: 031169 (8B26 06.35 Ashburton Grove-Blackbridge rubbish train and 8B35 08.42 return empties, with D8232)
Cambridge: 061169 (7P06 Welwyn Garden City-Whitemoor freight)
Welwyn Garden City: 211169 (additional Ashburton Grove-Blackbridge rubbish train and return empties, with D8232)
34D: 221169
Cambridge: 200170 (19.25 Palace Gates-Whitemoor freight)/ 070270 (N/B freight, with D5909)
Hitchin: 130270 (N/B freight)
34D: 140270

SFDRS: 150370/ 190370. SFDRSReps: 110370-200370 (radiators changed [stripped from D5902]).

Stevenage: 230470 (rescued Class 47 on "Hull Pullman" from Hitchin)
Biggleswade: 240470 (20.20 Edinburgh-KX, rescued failed loco. from Biggleswade)
Cambridge: 260570 (N/B engineer's train)
Chesterton Jct P.W. Depot: 260570
34G: 060670
Hitchin: 050770
Cambridge: 310770 (Hitchin-Chesterton Jct. l.e., with D5909; Chesterton Jct.-GN line ballast, with D5909)
34D Hitchin: 260970
34G: 251070
Hitchin: 021170 (09.15 Hitchin-Chesterton Jct ballast)
34D: 120271

Condemned: 1we200271 (140271) (rundown condition and broken auxiliary gearbox/fan shaft). Taken out of traffic 110271 (BDP).

34D: 210271/ 230271
34G Finsbury Park: 060371/070371
30A Stratford: Arrived 120371/ 140371 (DRS Erecting Shop)/ 270371/ 100471/ 170471/ 230571/ 300571/ 110671/ 100771 (inside New Shed)/ 310771/ 080871/ 280871 (New Shed Yard)/ 171071/ 281071/ 271171/ 191271/ 311271/ 030172 (New Shed Yard)/ 230172/ 050272/ 190272/ 120372/ 220472/ 200672/ 130772 (New Shed Yard)/ 270772/ 250872/ 030972/ 211072/ 181172/ 171272/ 311272/ 030273/ 250273/ 070373 (New Shed Yard)/ 250373/ 070473/ 230473/ 120573/ 140573/ xx0573 (adjacent to No.1 Diesel Shed ready for journey to Kettering). Power unit (engine+generator), traction motors and train-heating boiler removed; power unit retained at Stratford for D5901, if required; traction motors sent to Doncaster Works.

G. Cohen, Kettering: 300673/ 150773/ 120873/ 270873/ 060973 (intact)/ 190973. N/listed 270573/ 230673 and 240374.

Disposal: GCK 0673-1073 (D&ELfS).

Notes:
1. Mileage at 01/10/62 (still in traffic) since new: 133,210; mileage at 06/01/67 since refurbishment: 104,510.
2. The Stevenage Locomotive Society (*Aurora*) reported: 'The demise of the "Baby Deltics" was…..completed on this date [15 June 1963, the final Saturday of the winter timetable], D5905, after several weeks intermittent operation, being taken north by D5069.'
3. The Ipswich Transport Society (*Journal*, May 1971) reported D5905/9 as being 'inside the Diesel Repair Shop at Stratford at the end of March.'
4. Power unit (engine, main and auxiliary generators), traction motors and steam generator removed at Stratford following withdrawal as a source spares for D5901 in Departmental service.

D5906
EE/VF Works Nos.: 2383/D423

Above left: D5906, Hitchin, 15 January 1960. Radiator grille cover fitted. (Phillip Wells [RCTS Archive])

Above right: D5906, Stratford DRS, 24 August 1961. (Bill Hamilton [Grahame Wareham Collection])

Below left: D5906, King's Cross, 26 September 1966. (P. Chatman [Colour-Rail])

Below right: D5906, with D5907 in the distance, Doncaster Works (Erecting Shop Yard), 9 March 1969. Withdrawn and providing evidence that this locomotive never carried GFY livery as suggested in at least one magazine. (Adrian Booth)

New (date dispatched ex-EE): 080559 (with D5905).

Doncaster Works: 100559/ 120559 (main-line test run to New England with passenger stock, failed on arrival, hauled back to Doncaster by D5905 light-engines)/ 090659 (main-line test run). DLRC: 080569-110659 Trials.

To 34B: 110659

SFDRS: Nil. DLRC: 190859-210859 NC (air-brake modifications). N/listed 300859.

Holloway: 170959 (LE)

SFDRS: Nil. DLRC: 181159-281159 HC (Power-unit change). N/listed 101159.

Hitchin: 150160 (LE)

SFDRS: Nil. DLRC: 070360-080360 HC (Power-unit change). N/listed 270260.

34B: 120360
34G: 240460

SFDRS: Nil. DLRC: 190760-040860 HC (Power-unit change). N/listed 070760.

SFDRS: 181260/ 080161 (Works Yard)/ 150161. DLRC: 211260-280161 HI (No.2 Classified Repair & Modified Power-unit) (73,780 recorded miles).

Hertford North: 050361 (pass)
KX: 080361
Belle Isle: 180361 (N/B pass)
KX: 290361/ 130561 (pass)
XX: 220561 (15.57 Baldock-KX pass, failed with broken cardan shaft)
34G: 040661
Hitchin: 040861
Ganwick: 170861 (Cambridge-KX pass)

SFDRS: 240861 (Works Yard). DLRC: 210861-260861 HC (Power-unit change). N/listed 190861.

Belle Isle: 310861 (e.c.s.)
Hadley Wood: 121061 (pass)

SFDRS: Nil. DLRC: 251161-120162 HC (Power-unit change).

KX: 200362 (pass)/ 260362
XX: 260462 (St. Neots-Ferme Park freight, failed [wiring burnt out adjacent to boiler]) (BRCS)
34G: 290462

SFDRS: 270562. DLRC: 050562-310562 NC (boiler wiring renewed)

KX: 110762

SFDRS: Nil. DLRC: 210762-280762 NC (new radiators & fan gearbox repaired).

XX: 051062 (cracked liner injector hole) (DD)

SFDRS: Nil. N/listed 071062. DLRC: 151062-261162 (waiting engine).
Stored (u/s): 1we011262 (251162)
SFDRS/30A Stratford: 081262 (Works Yard)/ 030263/ 030363/ 060363/ 160363/ 310363/ 070463/ 210463/ 280463/ 250563/ 230663/ 070763

Re-instated: 3we030863 (280763).
Vulcan Foundry: Nil. DLRC: 290763-
Stored (u/s): 3we211263 (011263).
Vulcan Foundry: Nil. DLRC: 290763-xxxxxx.
Manchester Victoria: 161064 (with D6947)
Doncaster Works: 211064 (main-line test run), 231064 (main-line test run). DLRC: Doncaster Works: xxxxxx-291064 Trials.
Re-instated: 291064.

To 34G: 291064

SFDRS: 151164/ 221164 (Erecting Shop)

34D: 310165
Welwyn Garden City: 170465 (15.30 KX-Royston pass)
Stevenage: 040665 (13.09 KX-Cambridge 'Buffet' pass/ 15.15 Cambridge-KX 'Buffet' pass)
KX: 150665 (LE)
34G: 180765
KX: 140865
KX-Hitchin: 070965

GN Main-Line: 100965 (vans, with D1506)
XX: 301065 (21.25 Royston-Thames Haven empty oil tanks and return loaded tanks)

Newark: 011265 en route to Doncaster Works, with D5598)
36A Doncaster: 051265

XX: 281265 (fire damage; severe) (FDTL)

Works repair: ???

KX: 300466/ 060666 (N/B pass)

SFDRS: Nil. SFDRSReps: 050866 and 090866-240866 (No.2 Traction Motor).

KX-Potters Bar: 030966
KX s.p.: 090966
KX: 260966 (LE)
Hadley Wood: 211066 (pass)
34G: 231066
34D: 111266/ 090467/ 070567

Doncaster Works: 090767/ 270767/ 070867 (Erecting Shop)/ 230967/ 240967/ 011067

KX: 080368
Welwyn Garden City: 010568 (pass/e.c.s.)
KX: 150568

Stored u/s: xxxx68 (xx0568 per David Percival).
N.B. DLRC shows "Stored: ?-300968".

36A Doncaster: 020668/ 060668 (stored minus engine)/ 090668/ 160668/ 070768/ 140768/ 040868/ 010968/ 150968/ 190968/ 230968

Condemned: 1we051068 (300968).

36A Doncaster: 081068/ 151068/ 221068/ 291068/ 091168/ 121168
Doncaster Works: 011268 (Works Yard)/ 101268/ 050169/ 110169/ 210169/ 040269/ 250269/ 090369/ xx0369 (engine removed)/ 100469/ 170469 (Works Yard)/ 200469 (Works Yard)/ 040569/ 150669/ 120769/ 260769

Doncaster Belmont Yard: 090869 (with D5907)/ Departed 110869 (with D5907)

Chesterfield : xx0869 ("en route for Cohen's scrapyard, Kettering", with D5907)

Kettering Yard: 140869 ("waiting to come down branch to Cransley")
G. Cohen, Kettering: 070969/ 210969 (whole)/ 121069/ 241069/ 261069/ 311069/ 151169/ 241169/ 271169/ 111269 (whole)/ 261269/ 140170 (see note below).
N/listed 010969 and 270270/ 220470.

Disposal: GCK 0869-0771 (sic) (D&ELfS).

Notes:
1. Mileage at 01/10/62 (still in traffic [just]) since new: 145,810; mileage at 06/01/67 since refurbishment: 101,930.
2. The *Railway Observer* (May 1970) listed D5908 at G. Cohen, Kettering on 14 January 1970. D5908 was scrapped at J. Cashmore, Great Bridge and the RO reference in all probability related to D5906.

D5907
EE/VF Works Nos.: 2384/D424

Above left: D5907, Hatfield, Undated. (Rail-Online)

Above right: D5907, 30A Stratford, 28 April 1963. Stored pending refurbishment. (Brian Stephenson [Rail-Online])

Below left: D5907, King's Cross, 5 April 1965. Superb condition after only five days back in traffic after rehabilitation. (Rail-Online)

Below right: D5907 with D5906, Doncaster Works (Erecting Shop Yard), May 1969. Both withdrawn in GSY livery. (Sam Woods [David Dunn Collection])

New (date dispatched ex-EE): 150559.

Doncaster Works: 170559/ 200559 (main-line test run).
DLRC: 150569-220559 Trials.

To 34B: 220559
34B: 290559/ 060659
KX: 110759 (LE)
Hitchin: 120759 (08.33 Hitchin-Margate excursion, as far as ?)

SFDRS: Nil. DLRC: 040859-070859 NC (air-brake modifications).

Hitchin: 121059 (Cambridge Buffet Express)

SFDRS: Nil. DLRC: 181159-231159 HC (Power-unit change). N/listed 101159.

XX: 211259 (Moorgate-Hatfield pass)

SFDRS: Nil. DLRC: 260160-080260 HC (Power-unit change).

Cambridge: 260260
34B: 120360
KX: 260460/ 040660 ('Cambridge Buffet Express')

SFDRS: 260660. DLRC: 200660-210760 HC (Power-unit change).

34G: 310860
KX: 020960 (LE)

SFDRS: Nil. DLRC: 091160-031260 HC (Power-unit change).

SFDRS: 230461 (Erecting Shop)/ 300461 (Erecting Shop)/ 130561 (Erecting Shop). DLRC: 150461-130561 HI (No.2 Classified Repair) (90,590 miles recorded).

34G: 140561/ 040661
34D: 180661
Potters Bar: 220761 (pass)
Hitchin: 040861

SFDRS: Nil. DLRC: 290861-040961 HC (Power-unit change).

34D: 151061
Hadley Wood: 231061 (Pass)
34D: 230462
Hitchin: 120562 (engineer's train [track panels])

SFDRS: Nil. DLRC: 180562-240562 (traction motor cable repaired).

XX: 131162 (examination found metal deposits in 'chip trap', indicating advanced engine disintegration) (DD)

SFDRS: Nil. DLRC: 141162-261162 Waiting engine
Stored (u/s): 1we011262 (251162).
SFDRS/30A Stratford: 081262 (Works Yard)/ 030263/ 030363/ 060363/ 160363/ 310363/ 070463/ 210463/ 280463/ 250563/ 230663/ 070763. N/listed 070963.

Re-instated: 3we030863 (280763).
Vulcan Foundry: Nil. DLRC: 300763-
Stored (u/s): 3we211263 (011263).
Vulcan Foundry: Nil. DLRC: 300763-xxxxxx. Departed Vulcan Foundry 120365 (with D6971).
Penistone: 120365 (with D6971)
Doncaster Works: 130365/ 140365 (Paint Shop Yard)/ 170365 (main-line test run)/ 190365 (main-line test run).
DLRC: Doncaster Works: xxxxxx-310365 Trials.
Re-instated: 310365.

To 34G: 310365
KX: 050465 (LE)/ 150565
Stevenage: 040665 (13.53 Baldock-KX pass/ 16.22 KX-Royston pass pass/ 18.32 Royston-KX pass)
Hitchin: 100765 (pass)
KX: 100965
Knebworth: 281265 (17.15 KX-Royston pass)

SFDRS: Nil. SFDRS: 030166-100166 (radiators changed).

Doncaster Works: 070866. N/listed 200866.

Doncaster Works: Nil. N/listed 091066.
Grantham: 281066 (01.00 S/B with D5900)

34B: 111266
KX: 120167 (LE)
Hitchin: 080267 (e.c.s.)
34D: 090467
KX: 080767 (3E05 parcels)

34D: 050867
Moorgate: 140867 (08.53 arrival)
Cambridge: 180368 (pass)
KX: 230368 (stabled)/ 010568 (pass)
Moorgate: 060668 (08.54 arrival from Hertford North)

Doncaster Works: Nil. DLRC: xxxxxx-030768 U/C.

34D: 200768
KX: 130868
34G: 070968/ 080968

SFDRS: Nil. SFDRSReps: 110968-120968 (bogie wash).

Stored (u/s): 181068.
Condemned: 1we261068 (201068).

34G: 171168/ 071268/ 081268

Doncaster Works: 050169/ 110169/ 210169/ 040269/ 250269/ 090369/ xx0369 (engine removed)/ 100469/ 170469 (Works Yard)/ 200469 (Works Yard)/ 040569/ 150669/ 120769/ 260769

Doncaster Belmont Yard: 090869 (with D5906)/ Departed 110869 (with D5906)

Chesterfield : xx0869 ("en route for Cohen's scrapyard, Kettering", with D5906)

Kettering station: Nil. N/listed 140869. N.B. D5906 was present on this date.
G. Cohen, Kettering: Nil. N/listed 140869/ 010969/ 070969/ 210969/ 121069.

Disposal: GCK 0869 (D&ELfS).
Disposal not proven.

Notes:
1. Mileage at 01/10/62 (still in traffic) since new: 159,500; mileage at 06/01/67 since refurbishment: 89,330.
2. The Ipswich Transport Society (*Journal*, August 1968) indicated that D5907 was taken out of traffic around May/June 1968, along with D5906; however, a visit to Doncaster Works for an Unclassified repair in late June/ early July ensured a delay to official storage and withdrawal until October. Presumably, D5906's storage at 36A Doncaster at this time ensured the availability of spare parts.

D5907, Potters Bar, 22 July 1961. Southbound 'residential'. (Brian Stephenson [Rail-Online])

D5908

EE/VF Works Nos.: 2385/D425

Above left: D5908, King's Cross, 1960. (Alec Swain [Transport Treasury])

Above right: D5908, Location and date unknown. (Alec Swain [Transport Treasury])

Below left: D5908, 34G Finsbury Park, January 1968. After Unclassified Repairs and "touch up and varnish" treatment at Doncaster. (Rail-Online)

Below right: D5908, 34G Finsbury Park, 1969. GFY livery; double-arrow emblem. (Jim Blake [Grahame Wareham Collection])

New (date dispatched ex-EE): 290559.

Doncaster Works: 310559/ 030659 (main-line test run).
DLRC: 290559-050659 Trials.

To 34B: 050659
34B: 060659
Hadley Wood: 200659 (freight)
New Barnet: 280659 (pass)
Peterborough Westwood Goods Depot (exhibition): 110759/ 120759 (in failed condition?)

SFDRS: Nil. **DLRC:** 130759-140759 HC (Power-unit change by EE staff).

SFDRS: 230859. **DLRC:** 210859-240859 NC (air-brake modifications). N/listed 300859.

KX: 030959/ 071059 (pass)

SFDRS: 141159 (Works Yard). **DLRC:** 161159-191159 HC (Power-unit change). N/listed 101159.

SFDRS: Nil. **DLRC:** 090260-110260 HC (Power-unit change).

Wood Green: 220460 (pass)
KX: 260460

SFDRS: 010560. **DLRC:** 280460-050560 HC (Power-unit change).

SFDRS: 290560. **DLRC:** 300560-170660 HC (Power-unit change).

Oakleigh Park: 130960 (Hatfield-KX pass)

SFDRS: Nil. **DLRC:** 171060-291060 HC (Power-unit change).

34G: 121160
KX: 201260 (pass)

SFDRS: Nil. **DLRC:** 080261-130361 HI (No.2 Classified Repair) (80,070 miles recorded).

KX: 180361/ 200661 (LE)
Offord: 050761 (pass)

SFDRS: Nil. **DLRC:** 090861-120861 HC (Power-unit change).

Oakleigh Park: 160861 (Cambridge Buffet Express)
Huntingdon: 070961 (shunting pass stock)
KX: 011061/ 071061
New Southgate: 211061 (12.51 Biggleswade-KX pass)
34D: 151061

SFDRS: Nil. **DLRC:** 221161-011261 HC (Power-unit change).

XX: 010362 (06.43 Royston-KX pass, failed [loss of lubricating oil]) (BRCS)
34D: 230462
34G: 290462
Hitchin: 120562 (09:28 Ashburton Grove-New England freight)

XX: 230862 (defective turbo-charger) (DD)

SFDRS: 080962/ 061062/ 071062. **DLRC:** 290862-261162 (waiting engine).
Stored (u/s): 1we011262 (251162)
SFDRS/30A Stratford: 081262 (Works Yard)/ 030263/ 030363/ 060363/ 160363/ 310363/ 070463/ 210463/ 280463/ 230663/ 070763/ 200763
Cambridge: 200763 (hauled Stratford-Cambridge by D6711, with D5902)

Re-instated: 3we030863 (280763)
Vulcan Foundry: Nil. **DLRC:** 230763-
Stored (u/s): 3we211263 (011263)
Vulcan Foundry : Nil. **DLRC:** 230763-xxxxxx
Doncaster Works: 120764 (Paint Shop Yard)/ 200764 (main-line test run; failed at Newark (oil leaks), returned to Doncaster LE; stock returned by WD 90580)/ 270764/ 120864 (main-line test run). **DLRC:** Doncaster Works: xxxxxx-180864 Trials (defective traction motor replaced during trial period).
Reinstated: 180864

To 34G: 180864
Hitchin: 210864 (19.50 Royston-KX, replaced D5909 at Hitchin)
KX s.p.: 171064
XX: 301164 (07.35 Baldock-KX, failed with loss of lub oil)
Cambridge: 091264 (pass)

GN Main-Line: 040165 (assisting D1561 on up pass)
KX: 150565
Hitchin: 150665 (13.10 Baldock-KX pass)
34D: 030765
Cambridge: 310765 (pass)
Finsbury Park: 080965
Hitchin: 100965
St. Neots: 170965 (freight, replacement for D5673 requisitioned to assist an up pass)

Doncaster Works: 211165 (Works Yard)/ 051265/ 101265 (Erecting Shop). N/listed 141165 and 090166.

Doncaster Works: 170466 (Erecting Shop)/ 010566

Cambridge: 290866 ('Cambridge Buffet Express' to KX, with D5903)/ 300866 (ditto)/ 310866 (ditto)

Grantham: 301266 (en route Doncaster Works with D5586)
Doncaster Works: 080167

SFDRS: Nil. SFDRSReps: 020367-230367 (Fan gearbox [replacement gearbox from D5909]).

34D: 090467
34G: 070567
XX: 280667 (16.35 Finsbury Park CS-Broad Street e.c.s., failed, rescued by D5593)

Doncaster Works: 270767. N/listed 090767.

34D: 050867

Doncaster Works: 070867 (Works Yard). N/listed 230967.

Doncaster Works: 011067/ 151067. N/listed 240967.

XX: 301167. GFY livery, double-arrow emblem.
34G: 251267

Doncaster Works: 040168 (Works Yard)/ 070168 (Works Yard)/ 090168 (Erecting Shop). DLRC: xxxxxx-110168 U.

KX: 080568
XX: 130568 (ballast, with D5900)
Hertford North: 120668 (2B98 17.17 ex-Broad Street)
Langley Jct: 230668 (engineer's train)
34D: 200768
34G: 171168

Doncaster Works: 180169 (awaiting Works)/ 210169/ 230269. DLRC: 140169-xxxxxx U/C.

34G: 020369

Condemned: 1we150369 (090369).

34G: 220369
Ferme Park Down Sdgs: 090469
34B: 190469/ 200469
34G: 310569/ 080669/ 290669/ 050769/ 120769/ 260869/ 190969/ 031069 (departed)

Transfer 34G-Bescot: 031069

Bescot Yard/ 2F Bescot: xx1069 (arriving Bescot Yard hauled by Class 47)/ 061069 (arrived Bescot Yard)/ 271069 (Depot Yard)/ 091169/ 231169/ 291169 (Bescot Yard)/ 141269

J. Cashmore, Great Bridge: Arrived 050170 (Cashmore records)/ 080270/ xx0270 ("noted being cut-up")

Disposal: JCGB 0170 (D&ELfS).

Notes:
1. Mileage at 01/10/62 (stored by this date) since new: 141,580; mileage at 06/01/67 since refurbishment: 111,600.
2. **D5908** 'worked trials on numerous occasions' between 20/07/64 and 12/08/64. (BLS0964)
3. Auxiliary drive shaft and blower motor fitted to D5902 at Stratford DRS during period 200669-270669 using equipment stripped from D5908. This implies that D5908 briefly visited Stratford DRS at this time; the lack of an entry for D5908 in the SF DRS foreman/chargehand reports may reflect D5908 already being withdrawn.

D5909

EE/VF Works Nos.: 2386/D426

Above left: D5909, Stratford DRS, 20 May 1962. (Nigel Petre)

Above right: D5909, King's Cross, 12 June 1965. (Roger Norfolk)

Below left: D5909, King's Cross, 26 July 1965. (Stewart Blencowe)

Below right: D5909, 30A Stratford, 31 July 1971. The only recipient of BFY livery. (Rail-Online)

98 • ENGLISH ELECTRIC TYPE 2 BO-BO 'BABY DELTIC' LOCOMOTIVES

New (date dispatched ex-EE): 190659.

Doncaster Works: 210659/ 230659 (main-line test run)/ 100759/ 140759 (booked trials). Derived Record: Doncaster Works: 190659-xxxx59.

To 34B: xxxx59

SFDRS: Nil. BRDoc: 240859-260859 NC (air-brake modifications).

Skegness: 020959 (excursion)

SFDRS: Nil. BRDoc: 021259-181259 HC (Power-unit change).

SFDRS: Nil. BRDoc: 070360-080360 HC (Power-unit change).

34B: 120360

SFDRS: 270360 (Works Yard)/ 030460 (New Works). BRDoc: 280360-050460 HC (Power-unit change). N/ listed 260460.

SFDRS: 180960. BRDoc: 090860-190960 HC (Power-unit change).

KX: 261160 (pass)/ 101260

SFDRS: 150161/ 110261 (Erecting Shop). BRDoc: 160161-250261 HI (No.2 Classified Repair).

34D: 180661
Biggleswade: 230761 (Biggleswade-Mablethorpe excursion)
Hitchin: 040861

SFDRS: 071061. BRDoc: 260961-201061 HC (Power-unit change).

SFDRS: 051161 (Erecting Shop). BRDoc: 011161-071161 HC (lifted for new axlebox fitting).

34D: 051261 (exam, cracked cylinder liner) (BRCS)

SFDRS: 070162/ 210162/ 250262 (Erecting Shop). BRDoc: 081261-030362 HC (Power-unit change).

XX: 070362 (17.48 KX-Welwyn Garden City, failed [con rod through crankcase]) (BRCS)

SFDRS/30A Stratford: 080462 (Works Yard)/ 150462 (Erecting Shop)/ 290462/ 200562/ 270562/ 070662/ 260762/ 280762/ 290762/ 120862 (Works Yard)/ 260862 (Works Yard)/ 080962/ 061062/ 071062/ 211062/ 041162

Stored (u/s): we011262 (251162).
SFDRS/30A Stratford: 081262 (Works Yard)/ 030263/ 030363/ 060363/ 160363/ 310363/ 070463/ 210463/ 280463/ 230663/ 070763.

Re-instated: 3we080363 (280763).
Vulcan Foundry: Nil.
Stored (u/s): 3we211263 (011263).
Vulcan Foundry: Nil.
XX (Manchester area): 030764
Doncaster Works: 050764 (Paint Shop Yard)/ 070764 (main-line test run)/ 120764 (Paint Shop Yard). Derived Record: xxxxxx-160764 Trials.
Re-instated: 160764.

To 34G: 160764
KX: 170764
Stevenage: 100864 (failed on 05.22 Hitchin-KX with loss of lub oil)
Hitchin: 210864 (19.50 Royston-KX, replaced by D5908 at Hitchin)

XX: 03-040964 (radiators changed).

Finsbury Park-Cambridge: 090964
KX: 100964
KX-Cambridge: 110964
34G: 260964
KX: 041064

Newark: 091164 (en route to Doncaster Works)
Doncaster Works: Arrived 091164 (oil leaks)

XX: 301164 (05.25 Hitchin-Sandy e.c.s., failed with loss of lub oil)
Cambridge: 071264 (pass)
Hatfield: 040165 (07.35 Baldock-KX, failed at Hatfield with loss of lub oil, assisted by D8046 to KX)
GN Main-Line: 220465 (up cement, with D6507)
KX: 150565

Stevenage: 040665 (08.24 KX-Baldock pass/ 17.36 FO Huntingdon-KX)
KX: 120665
34D: 030765
34G: 180765
KX: 260765 (N/B pass)
Hitchin: 310865 (S/B pass)/ 070965
34B: 160166

Doncaster Works: 111266. Released with overhauled engine No.388 (BDP).
Grantham: 151266 (assisting failed D1513 on 1A09 Hull-KX pass, from Retford)

KX: 291266

SFDRS: Nil. SFDRSReps: 210267-100467 (radiators changed).

KX: 090767 (S/B pass)
Potters Bar: 130867 (engineer's train)
Moorgate: 231067 (07.56 ex-Hertford North)

Doncaster Works: Nil. Engine No. 388 failed on 111267 with piston defect; engine removed at Doncaster Works for repairs by Napier; engine subsequently re-installed in D5905 (BDP).
Doncaster Works: 040168 (Erecting Shop)/ 070168 (Erecting Shop)/ 090168 (Erecting Shop)/ 170268/ 190268 BFY livery (Station Yard). Ex-Works: 160268 (DD No.29). N/listed 280168/ 040268 and 030368.

34G: 030468
KX: 150568
Stevenage: 290568 (19.20 Palace Gates-Whitemoor coal empties, with D5900)
Langley Jct: 310568 (Palace Gates-Whitemoor coal empties, with D5626)/ 130668 (Palace Gates-Whitemoor coal empties, with D5900)
Hertford North: 120668 (pass)
34D: 200768
KX: 010868
Welwyn Garden City/Knebworth: 130868 (exhaust fire)
KX: 180968
Woolmer Green: 151168 (07.15 Royston-KX pass; caught fire, extinguished by driver)
34G: 171168

Doncaster Works: 261168/ 011268 (Works Yard)/ 101268

SFDRS: 191268/ 241268 (DRS Yard)/ 251268. SFDRS: 181268-241268 (blower fan drive [replacement drive ex-D5900 fitted]).

34D: 180169
KX: 070269 (LE)
34D: 230269
KX: 180369 (pass/e.c.s.)
Stevenage: 020469 (pass)
KX: 100469/ 140469 (N/B pass)
34G: 200469
34D: 250469
Cambridge: 060669 (hauled failed D5902 from Cambridge to Hitchin)
34D: 080669/ 270669
KX: 120769
Harringay West: 070869 (pass)
KX: 150869 (17.39 KX-Peterborough pass, in lieu of Class 47, with D5905)
34D: 180869/ 190969
Welwyn Garden City: 190969 (08.43 King's Cross-Baldock pass)
GN Main-Line: 241069 (freight)
KX: 041169 (arrived piloting D9017 on 1E48)
Cambridge: 141169 (13.04 KX-Cambridge pass)
34D: 161169
34G: 231169
Offord: 300170 (engineer's train)
Cambridge: 070270 (N/B freight, with D5905)/ 100270 (rails to Chesterton Jct.)
34D: 140270
KX: 200270 (e.c.s. to Moorgate for 16.58 Moorgate-Hertford North)
34D: 070370
Ely: 140370 (freight from Whitemoor to GN line)
Holloway: 190370 (engineer's train)
34G: 200470
Cambridge: 210470 (7P06 Welwyn Garden City-Whitemoor freight)/ 230470 (ditto)
Cambridge Yard: 020570 (stabled on breakdown train)
34D: 070670/ 020770
Hitchin: 050770
Cambridge: 210770 (Welwyn Garden City-Whitemoor freight)/ 310770 (Hitchin-Chesterton Jct. l.e., with D5905; Chesterton Jct.-GN line ballast, with D5905)
Hitchin: 260970 (engineer's train)
Baldock: 101170 (fire damage, not severe; hauled back to Hitchin for repairs) (FDTL)

GN Main-Line: 161170 (19.31 Welwyn Garden City-Whitemoor freight, 22.15 Whitemoor-KX freight)
Cambridge: 181270 (hauled failed D6743 from GN line to 31B March)
34D: 120271/ 210271/ 230271
34G: 060371/ 070371

Condemned: 1we130371 (070371) (rundown condition).

34G: 180371
30A Stratford: 270371/ 170471/ 090571 (alongside New Shed)/ 110571/ 230571/ 300571/ 110671/ 100771 (inside New Shed)/ 310771/ 080871/ 280871 (New Shed Yard)/ 171071/ 291071/ 271171/ 191271/ 311271/ 030172 (New Shed Yard)/ 230172/ 050272/ 190272/ 120372/ 220472/ 290472/ 200672/ 130772 (New Shed Yard)/ 270772/ 250872/ 030972/ 211072/ 181172/ 171272/ 311272/ 030273/ 250273/ 070373 (New Shed Yard)/ 250373/ 070473/ 230473/ 120573/ 140573/ xx0573 (adjacent to No.1 Diesel Shed ready for journey to Kettering). Power unit (engine + generator) removed (destination unknown); traction motors removed and sent to Doncaster Works.

G. Cohen, Kettering: 300673/ 150773/ 120873/ 270873/ 060973 (intact)/ 190973/ 020274/ 240374/ 050774 (gutted internally, part-cut). N/listed 270573/ 230673 and 170275/ 220275.

Disposal: GCK 0673-1073 (D&ELfS).

Notes:
1. Mileage at 01/10/62 (stored by this date) since new: 106,350; mileage at 06/01/67 since refurbishment: 125,510.
2. The Ipswich Transport Society (*Journal*, May 1971) reported D5905/9 as being 'inside the Diesel Repair Shop at Stratford at the end of March'.

D5909, 30A Stratford, May 1973. Positioned adjacent to 'A' shed with D5905 prior to movement to Cohen's scrap yard at Kettering for disposal. (Rail-Photoprints)

Chapter 9
PERFORMANCE AND SERVICE PROBLEMS (1959–63)

9.1 Early Performance – Summary

The 'Baby Deltics' initially proved to be very satisfactory in service but a range of problems developed as the months passed to the extent that by November 1959 seven locomotives had already had their engines changed.

The weekly 'Baby Deltic' availability figures during the eleven weeks ending 22 October 1960 only exceeded sixty per cent once with five of those weeks at fifty per cent. These figures are misleading, however, in the sense that locomotives had deliberately been taken out of service for campaign engine modifications; interestingly, a BR document dated 25 January 1960 indicated that a planned modification programme was due to start in February 1960, but this was subsequently delayed due to the ongoing need for emergency repairs to 'Deltic' engines to keep the fleet in service. BR's availability return for the four week period ending 10 September 1960 specifically mentioned 'programmed modifications' (see Section 9.2.1).

It is possible that the Works visits for these modifications were effectively No.1 Heavy Intermediate repairs, although they were only ever described as Heavy Casual repairs on the Diesel Locomotive Record Cards (DLRC). Modification dates are believed to be as follows (as selected from available DLRCs and King's Cross District reports and based on the fact that modifications were scheduled to take 10-18 days):

Loco. No.	Scheduled Works Dates (as at 25/01/60)	Actual Works Dates
D5900	20/04/60-05/05/60	13/09/60-07/10/60
D5901	06/03/60-23/03/60	20/05/60-01/06/60
D5902	20/04/60-05/05/60	17/05/60-27/05/60, or, 04/08/60-14/09/60
D5903	06/03/60-23/03/60	04/08/60-24/09/60
D5904	26/03/60-12/04/60	09/06/60-25/06/60
D5905	08/03/60-25/03/60	23/06/60-29/07/60
D5906	04/03/60-21/03/60	19/07/60-04/08/60
D5907	29/02/60-17/03/60	20/06/60-21/07/60
D5908	17/02/60-05/03/60	30/05/60-17/06/60
D5909	19/03/60-05/04/60	09/08/60-29/09/60

By October 1960, part way through the modification programme, the 'Baby Deltics' had run 40-60,000 miles each, adversely affected by equipment failures, and since the first locomotives entered service in April 1959 the number of engine changes carried out at Stratford Works had increased to forty-four. Some of the identified causes were:

1. Fractured cylinder liners — 10
2. Piston failures — 5
3. Turbo-charger turbine failures — 10
4. Engine de-phased — 1
5. Miscellaneous faults — 4
6. Scheduled engine modifications — 7

The main areas of focus of the modification programme are clearly evident. Engine cooling difficulties also experienced during 1960 resulted in a temporary de-rating of the engine to 1025hp and as reported by *Trains Illustrated* (November 1960) the idea of transferring the 'Baby Deltics' away from the King's Cross area was apparently mooted:

> It is expected that the English Electric Type 2s, now de-rated to 1025h.p., will be transferred away from the Finsbury Park depot, but their next field of operation has yet to be decided.

The *Railway Observer* (September 1960) suggested Sheffield colliery work. The only other information on this idea surfaced in Robert Stephens' book *Diesel Pioneers* (1988) in which he states:

> After tests were conducted on 12th February 1960 when D5903/5 worked in multiple on a coal train from Whitemoor to Bury St. Edmunds, it was planned to allocate the class to the Sheffield area.

Nothing came of that idea and the 'Baby Deltics' were returned to 1,100hp following radiator fan and grille modifications, presumably undertaken during the August-October 1960 modification period.

Even before completion of the first wave of modifications, plans were afoot in July 1960 to further modify the 'Deltic' T9 engines with the aim of obtaining an operating period of 4,000 hours between engine changing. Napier's plans to introduce a programme to completely rebuild the 'Deltic' engines, incorporating all necessary modifications, was accepted by BR and the first rebuilt ones were due for delivery in December 1960. The so-called 'zero-life' modifications were undertaken as part of a No.2 Heavy Intermediate repair and the engines were wound back to zero hours. Modifications included:

- fully modified cylinder liners,
- new pistons in cases where incorrect lubricating oil had damaged the pistons,
- modification of the top crankcases with stiffer webs,
- provision of an increased bearing area for the turbo-charger compressor impeller bearings and change of material from aluminium to steel,
- fitting of new seals on the coolant pump and provision of increased clearances in the thermostatic valves, and,
- modifying the engine governor with spring links on the control linkage.

The first modified locomotive actually returned to traffic in January 1961. Full details are:

Loco. No.	"Zero-Life" Programme Modifications	Comments	Mileage at No.2 Intermediate Repair
D5900	18/01/61-04/03/61		75380
D5901	24/02/61-08/04/61		
D5902	09/02/61-25/03/61		
D5903	05/01/61-17/02/61		70750
D5904	10/03/61-20/04/61		75700
D5905	25/03/61-29/04/61		
D5906	21/12/60-28/01/61	First loco	73780
D5907	15/04/61-13/05/61	Last loco	90590
D5908	08/02/61-13/03/61		80070
D5909	16/01/61-25/02/61		

The plan was that following the 'zero-life' modification programme locomotives would only be returned to Stratford Works for interim 2000/2250hr examinations.

Availability during four weeks of the first sixteen weeks of 1961 was 50 per cent or below, in part attributable to locomotives being in Works for modifications. Improvements in the availability statistics were evident from May 1961 as a consequence of Napier's modifications but, despite this attention, availability started to fall away once again from December 1961, with instances of 40 and 50 per cent weekly availability being reported in 1962. This situation made locomotive rostering an impossible task and by late 1961/early 1962, BR was losing faith in the 'Baby Deltics', particularly given that other Type 2s (Brush and BR/Sulzer) were proving to be more reliable.

Locomotives were progressively taken out of use following major failures from February 1962 pending resolution of the issues. The last repair at Stratford Works appears to have been D5905 in May 1962.

In May 1962 English Electric published another report which again dealt with 'Deltic' engine problems. Based on an analysis of engine failures, it was stated that, with some further modifications, the engine could be made reliable. A survey of 24 engine failures revealed that thirteen were due to fractured cylinder liners, two to defective liner sealing rings, plus one each for output bearing and auxiliary generator drive issues; five were due to external causes such as freezing-up, etc., with two cases still under investigation.

Over a period of sixteen months up to June 1963, all ten 'Baby Deltics' removed themselves from traffic by succumbing to engine failure. BR Period Reports started including references to the possibility of replacing the 'Deltic' engine with the first comment made in the return for Period Ending 16 June 1962.

9.2 Performance Statistics

9.2.1 BTC/BRB Locomotive Performance Reports (1959-63)

The monthly minutes of the BTC Motive Power Committee included an Appendix A entitled 'Diesel Main-Line Locomotives Allocation and Performance'; records have been found covering the period August 1959-December 1962. In addition, BRB operational reports provided further statistics up to July 1963. 1959–61 results were reported weekly; however, 1962 and 1963 results were, in the main, reported for the last week of each 4-week administrative period only.

The statistics listed below cover the performance aspects of the 'Baby Deltic' fleet (i.e. reliability and utilisation); to save space, the 1959–62 weekly results are aggregated to monthly statistics (weighted to take into account the number of locomotives in stock each week).

Performance Report (1959-62)

Period-Ending (P/e), or, Week-Ending (W/e)	Operating Stock	Availability % Av. Whole 24 hours	Miles per Loco per Weekday	
			In Use	Stock
P/e 08/08/59	10	75	169	122
P/e 05/09/59	10	75	180	130
P/e 03/10/59	10	88	178	147
P/e 31/10/59	10	90	168	150
P/e 28/11/59	10	78	175	135
P/e 26/12/59	10	75	176	130
P/e 30/01/60	10	80	172	134
P/e 27/02/60	10	85	175	144
P/e 26/03/60	10	88	185	153
P/e 23/04/60	10	90	184	160

ENGLISH ELECTRIC TYPE 2 BO-BO 'BABY DELTIC' LOCOMOTIVES

Period-Ending (P/e), or, Week-Ending (W/e)	Operating Stock	Availability % Av. Whole 24 hours	Miles per Loco per Weekday	
			In Use	Stock
P/e 21/05/60	10	78	182	141
P/e 18/06/60	10	78	173	126
P/e 16/07/60	10	78	180	139
P/e 13/08/60	10	70	197	132
P/e 10/09/60	10	58	181	104
P/e 08/10/60	10	63	147	88
P/e 05/11/60	10	63	150	90
P/e 03/12/60	10	68	164	104
P/e 31/12/60	10	88	157	134
P/e 28/01/61	10	60	162	94
P/e 25/02/61	10	55	158	83
P/e 25/03/61	10	58	158	91
P/e 22/04/61	10	68	179	113
P/e 20/05/61	10	83	150	123
P/e 17/06/61	10	88	203	167
P/e 15/07/61	10	90	186	152
P/e 12/08/61	10	95	180	156
P/e 09/09/61	10	88	181	154
P/e 07/10/61	10	93	165	153
P/e 04/11/61	10	78	203	154
P/e 02/12/61	10	85	199	167
P/e 30/12/61	10	55	193	109
W/e 27/01/62	10	60	131	79
W/e 24/02/62	10	40	248	99
W/e 24/03/62	10	40	188	75
W/e 21/04/62	10	50	153	76
W/e 19/05/62	10	40	148	44
W/e 16/06/62	10	50	154	46
W/e 14/07/62	10	60	182	73
W/e 11/08/62	10	60	188	38
W/e 08/09/62	10	40	132	53
W/e 06/10/62	10	20	210	42

Performance and Service Problems (1959–63) • 105

Supporting commentary was provided including factors affecting availability and are provided in abridged form below:

Period Ending	Comments
08/08/59 & 05/09/59	Modifications to brakes and radiators.
03/10/59	Improvement in availability compared with previous period.
31/10/59	Availability above average for we 17/10/59.
28/11/59 & 26/12/59	Main Works attention to turbo-blower bearings.
30/01/60	Locos in Main Works for engine changing due to failure of turbo-blower.
27/02/60	Availability reasonably good.
26/03/60	Availability reasonable.
23/04/60 & 21/05/60	No comments.
18/06/60 & 16/07/60	Locos in Main Works for engine changing due to piston and cylinder liner defects.
13/08/60	Continued piston and cylinder liner troubles being experienced.
10/09/60	Increased number of locos in Main Works for programmed modifications.
08/10/60	Locos in Main Works for engine changing as a result of piston and cylinder liner failures.
05/11/60	Locos in Main Works awaiting reconditioned engines from the manufacturer.
03/12/60	Locos awaiting reconditioned engines from the manufacturer.
31/12/60	Agreed changing of engines has affected availability.
28/01/61	Locos stopped in and awaiting Works due to engine changes.
25/02/61	Practically 50% of locos have been in Main Works for engine changes.
25/03/61	Engine changing.
22/04/61	Second intermediate repairs and engine changes.
20/05/61	Two locos in Works for intermediate repairs and one under repair at depot during we 29/04/61.
17/06/61	One loco in Main Works for attention.
15/07/61	Performance satisfactory.
12/08/61	Reasonable performance.
09/09/61	Cracked cylinder liners and unit changes at Main Works.
07/10/61	One loco undergoing engine repairs at Main Works.
04/11/61	Power unit changes to three locos, plus other casual heavy repairs.
02/12/61	Power unit changes.
30/12/61	Cracked cylinder liners and other engine failures resulted in locos being taken into Works for power unit changes. At the end of the period five locos awaiting replacement units from manufacturer.
27/01/62	Three out of four locos out of traffic awaiting replacement power units at Main Works.
24/02/62	Power unit problems caused by fractured liners; six locos awaiting replacement power units.
24/03/62	Five locos in Works for power unit changes and one stopped at depot (faulty AWS).
21/04/62	Power unit changes due to fractured liners accounted for all locos not available.

Period Ending	Comments
19/05/62	Poor availability due to four locos undergoing power unit changes due to fractured liners and one loco undergoing traction motor cable repairs in Main Works.
16/06/62	Four locos in Main Works for power unit changes due to fractured liners, one loco at depot for examination. The future of this type of engine is to be decided, pending this repairs to the engine are in abeyance.
14/07/62	Poor availability due to four locos in Main Works awaiting change of engines from manufacturers.
11/08/62	Four locos in Works awaiting change of engines. Questions of changing 'Deltic' (*engine*) for one of another type in hand with BTC. Locomotives in traffic used on light work.
08/09/62	Poor availability due to four locos awaiting replacement power units and one loco with turbine failure in Main Works. One loco in depot for attention by makers. The question of changing this type of power unit still under discussion.
06/10/62	Five locos in main Works awaiting replacement power units. One loco awaiting repairs. two locos for examination at depots. Question of change of engine still under review.

9.2.2 Monthly Report to BTC

Information from the Monthly Reports to the BTC/BRB entitled 'Diesel Train Locomotives: Availability and Utilisation' (from Period 10 1962 to Period 8 1963):

Performance Report (1962-63).

| Week-Ending (W/e) | Net. Operating Stock | Actual | | | | Diagram | | | Remarks |
| | | Av. No. Available (Weekday) | | Av. Miles Worked per Weekday per Loco | | Av. No. required each Weekday to work all Diagrams | Av. Diagram Miles per Weekday per Loco | | |
		Used	Not Used	In Stock	In Use		In stock	Required	
W/e 06/10/62	10	2	-	42	210	4	119	297	93% passenger
W/e 03/11/62	10	3	1	53	175	4	120	297	93% passenger
W/e 01/12/62	3	2	-	110	165	4	369	297	93% Passenger
W/e 22/12/62	3	3	-	127	127	4	369	297	93% passenger
W/e 26/01/63	3	2	-	90	134	-	-	-	N/D
W/e 23/02/63	1	1	-	117	117	-	-	-	N/D
W/e 23/03/63	2	1	-	59	119	-	-	-	N/D
W/e 06/04/63	2	1	-	35	70	-	-	-	N/D
W/e 18/05/63	2	1	-	68	136	-	-	-	N/D
W/e 15/06/63	2	1	-	27	53	-	-	-	N/D
W/e 13/07/63	1	-	-	-	-	-	-	-	In Works
W/e 10/08/63	10	-	-	-	-	-	-	-	In Works

Note:
N/D = Not diagrammed; used as required.

9.2.3 'Availability by Classes of Locomotives and Power Cars' (BR.1712/444)

Eastern Region

S(u)	Stored unserviceable.
Net Op. Stock	Net Operating Stock
Not Available Works	In Works & Awaiting Works
Not Available Depots	Under or Awaiting Repairs and Under or Awaiting Exams

Average Weekday Position (1962)
Number and Percentage of Net Operating Stock

Availability (1962).

Week-Ending (W/e)	S(u)	Net. Op. Stock	Available No. (%)	Not Available	
				Works No. (%)	Depots No. (%)
W/e 17/02/62	-	10	5 (50)	4 (40)	1 (10)
W/e 24/02/62	-	10	4 (40)	5 (50)	1 (10)
W/e 03/03/62	-	10	5 (50)	4 (40)	1 (10)
W/e 10/03/62	-	10	4 (40)	4 (40)	2 (20)
W/e 17/03/62	-	10	5 (50)	5 (50)	0 (0)
W/e 24/03/62	-	10	4 (40)	5 (50)	1 (10)
W/e 31/03/62	-	10	3 (30)	6 (60)	1 (10)
W/e 07/04/62	-	10	5 (50)	5 (50)	0 (0)
W/e 14/04/62	-	10	6 (60)	4 (40)	0 (0)
W/e 21/04/62					
W/e 28/04/62	-	10	5 (50)	4 (40)	1 (10)
W/e 05/05/62					
W/e 12/05/62	-	10	3 (30)	7 (70)	0 (0)
W/e 19/05/62	-	10	4 (40)	5 (50)	1 (10)
W/e 26/05/62	-	10	4 (40)	5 (50)	1 (10)
W/e 02/06/62	-	10	5 (50)	4 (40)	1 (10)
W/e 09/06/62	-	10	5 (50)	4 (40)	1 (10)
W/e 16/06/62	-	10	5 (50)	4 (40)	1 (10)
W/e 23/06/62	-	10	6 (60)	4 (40)	0 (0)
W/e 30/06/62	-	10	6 (60)	4 (40)	0 (0)

Week-Ending (W/e)	S(u)	Net. Op. Stock	Available No. (%)	Not Available Works No. (%)	Depots No. (%)
W/e 07/07/62	-	10	6 (60)	4 (40)	0 (0)
W/e 14/07/62	-	10	6 (60)	4 (40)	0 (0)
W/e 21/07/62	-	10	4 (40)	5 (50)	1 (10)
W/e 28/07/62	-	10	4 (40)	5 (50)	1 (10)
W/e 04/08/62	-	10	5 (50)	5 (50)	0 (0)
W/e 11/08/62	-	10	6 (60)	4 (40)	0 (0)
W/e 18/08/62	-	10	6 (60)	4 (40)	0 (0)
W/e 25/08/62	-	10	5 (50)	4 (40)	1 (10)
W/e 01/09/62	-	10	4 (40)	5 (50)	1 (10)
W/e 08/09/62	-	10	4 (40)	5 (50)	1 (10)
W/e 15/09/62	-	10	5 (50)	5 (50)	0 (0)
W/e 22/09/62	-	10	5 (50)	5 (50)	0 (0)
W/e 29/09/62					
W/e 06/10/62	-	10	2 (20)	5 (50)	3 (30)
W/e 13/10/62	-	10	3 (30)	6 (60)	1 (10)
W/e 20/10/62	-	10	3 (30)	6 (60)	1 (10)
W/e 27/10/62	-	10	4 (40)	6 (60)	0 (0)
W/e 03/11/62	-	10	4 (40)	6 (60)	0 (0)
W/e 10/11/62	-	10	3 (30)	6 (60)	1 (10)
W/e 17/11/62	-	10	2 (20)	7 (70)	1 (10)
W/e 24/11/62	-	10	2 (20)	7 (70)	1 (10)
W/e 01/12/62	7	3	2 (67)	0 (0)	1 (33)
W/e 08/12/62	7	3	3 (100)	0 (0)	0 (0)
W/e 15/12/62	7	3	3 (100)	0 (0)	0 (0)
W/e 22/12/62	7	3	3 (100)	0 (0)	0 (0)
W/e 29/12/62	7	3	3 (100)	0 (0)	0 (0)

Average Weekday Position (1963) (N.B. Selected weeks only reported {usually last week of four-week period)).
Number and Percentage of Net Operating Stock

Availability (1963).

Week-Ending (W/e)	S(u)	Net. Op. Stock	Available	Not Available	
				Works	Depots
			No. (%)	No. (%)	No. (%)
W/e 26/01/63	7	3	2 (67)	0 (0)	1 (33)
W/e 23/02/63	9	1	1 (100)	0 (0)	0 (0)
W/e 23/03/63	8	2	1 (50)	1 (50)	0 (0)
W/e 06/04/63	8	2	1 (50)	1 (50)	0 (0)
W/e 18/05/63	8	2	1 (50)	1 (50)	0 (0)
W/e 15/06/63	8	2	1 (50)	1 (50)	0 (0)
W/e 13/07/63	9	1	0 (0)	1 (100)	0 (0)
W/e 03/08/63	-	10	0 (0)	10 (100)	0 (0)

9.3 Mechanical Issues

Early mechanical problems experienced with the 'Baby Deltics' included:

- Poorly seated injector adaptors (seven engines had received injector modifications by end-August1959).
- Cylinder liner cracks radiating from the holes into which the liner locating adaptor plug was screwed (by early October 1959 five engines had been changed because of fractured liners).
- Turbo-charger turbine bearings failures, with four failures occurring during November 1959 alone; the failures all occurred at a similar number of service hours enabling preventative action to be taken as other engines approached those hours.
- Leaking high-pressure fuel pipes, rectified in the short-term by tightening joints or by the renewal of pipes.
- Broken lubricating oil pressure pump quill shafts.
- Engine shutdowns due to low water level in the header tank due to the coolant being forced under pressure back into the header tank and out of the overflow. In most cases this fault did not prevent restarting of engines because after shutdown much of the coolant drained back into the engine.
- Auxiliary gearbox failures and associated issues.

Some of these issues were resolved relatively quickly but a number remained persistent problems right up to the point that the locomotives were taken out of service in 1962/63 pending a decision as to their future. The most persistent problems are covered in the following sections.

9.3.1 'Deltic' T9 Engine – Fractured cylinder liners

Fractured cylinder liners on the 'Baby Deltics' were a relatively common occurrence with a number of causes:

- Stresses set up during assembly of the engines. Tightening of the crankcase bolts caused the cylinder block to bear hard against the injector adaptor, itself screwed into the cylinder liner. This area of local stress led to liner fractures.
- Erosion caused by electrolytic action in the threaded injector adaptor hole.

'Recuttering' of the adaptor holes was undertaken at Finsbury Park with plating of the holes with lead, tin or gold to prevent erosion. However, initial efforts to prevent erosion failed and improved liners were designed, incorporating thicker walls, better securing methods and improved cooling.

Cylinder liner fractures were also a persistent problem at the injector opening which, being at the mid-point of the liner, was the most highly loaded portion having to bear the full firing pressure until the two opposing pistons began to move apart. In four-stroke engines there was some 'give' in the cylinder head studs, but in the opposed-piston engine there was nothing to 'give' apart from the liner. The potential for strengthening in this area was limited by the need for water cooling galleries to remove heat.

Cavitation erosion of the alloy cylinder block occurred in the coolant space adjacent to the cylinder liner scrolls due to untreated coolant water being used; specially treated coolant water proved to be absolutely essential requirement with 'Deltic' engines.

9.3.2 'Deltic' T9 Engine – Piston Failures

Piston seizures due to the chemical erosion of the underside of the piston crown occurred due to heat transfer issues. In opposed-piston two-stroke engines the piston, and in particular the top piston ring, had to perform the double function of acting as a piston ring and as a valve. This was particularly hard on the exhaust piston and resulted in a considerably greater heat dissipation requirement than in a comparable piston in a four-stroke engine.

Although not directly responsible for engine failures in traffic, the loosening of the 'Hydural' piston crowns was a troublesome feature. Modified fits and increased tightening were tried early on but were not entirely successful in eliminating the problem. The design went through several iterations before a satisfactory crown was produced, only achieved after the T9 engine rehabilitation modifications in 1963/64.

9.3.3 Turbo-Charger Failures

Turbo-charger turbine bearing failures were caused by back pressure in the locomotive exhaust system. Exhaust gases leaked past the turbine oil labyrinth seals, forming carbon deposits on the turbine rotor and causing a heavy end thrust on the bearings which they were not designed to withstand. Seals, air delivery, and routing were modified.

9.3.4 Cooling System Inadequacies

Persistent problems were experienced with the cooling system, specifically:

- Oil and water leakages due to the weakness of radiator/pipework soldered joints, radiator matrix gasket deterioration (particularly in the water section of the radiator), and, radiator matrix leaks;
- Oil and water leakages due to inadequate circuit pipework and jointing (particularly in the lubricating oil circuit due to corrosion issues and variable pressures);
- Overheating (due to blocked radiators, incorrect water treatment, incorrectly set thermostats, poor radiator cleaning externally and internally, etc.).

Leaking radiator/pipework joint connection problems were resolved to a degree by the use of alternative soldering materials and modified connections. Blame was initially placed on differential thermal expansion within the radiator frame/matrix interfaces but Marston Excelsior, the radiator manufacturer, blamed the leakages on vibration from the locomotives' superstructure transmitted to the radiator by too-rigid pipework causing the radiator joints to fail; despite extensive laboratory tests to replicate service conditions, excessive vibration was not demonstrated to be an issue.

Solutions deployed in an attempt to combat the problems generated in the challenging traction environment included:

- A revised oil radiator header design, including enhanced structural arrangements to mitigate against differential expansion and vibration (there was a proposal to modify the water radiator design similarly but it is not known if this re-design took place);
- Rubber mountings in the radiator frame;
- Convoluted hoses at critical points instead of rigid pipe joints (instigated at Marston's request).

A revised radiator matrix gasket material was introduced to prevent deterioration and resolution of radiator leaks was attempted by the use of alternative coating materials.

Pipework joint leakages were never fully resolved but various initiatives were introduced to improve the situation:

- The corrosive effects (dezincification of pipes) of lubricating oil was reduced by the use of oil from an alternate supplier;
- Revised instructions regarding the fitting and subsequent inspection of bolts.

As already mentioned, engine cooling deficiencies (i.e. overheating) resulted in a temporary de-rating of the 'Deltic' engines from 1,100hp to 1025hp in 1960. Various modifications were made during the mid-1960s to obviate the problem and these proved to be effective, including:

- Redesigned and modified air-ducting;
- Fitting 12-bladed, instead of 8-bladed, cooling fans to increase airflow past the radiators to meet all contingencies;
- Removal of the two external bodyside radiator grille covers again to improve air flow;
- Implementation of an improved external radiator cleaning regime.

Lubricating oil was cooled in its own section of the radiator. Burst lubricating oil hoses were found to correlate with very high oil pressures emanating from the scavenge oil pump following early morning cold starts and as these locomotives were regularly rostered on suburban duties with no night-time activity, they frequently had to start from cold each day. The pressure-rise as a consequence of cold viscous oil was considered to be so extreme and rapid that the flexible hoses burst before the relief valve opened; however, laboratory tests undertaken using more extreme pressure fluctuations than that found in traffic failed to result in any joint failures.

Various radiator tests involving Marston, Napier and BR were undertaken during 1960 and into 1961 but, it would appear from contemporary literature, without ultimately obtaining a full scientific understanding of the causes of the problem and, therefore, without deriving an effective resolution of the problem. Various modifications were made over time in the hope of achieving some improvement if not a complete resolution.

Problems were sufficiently serious for an internal EE memo author to comment in January 1961, '… confidence (in Marston Excelsior) must be impaired in a product which had occasioned this disproportionate history of fault.'

9.3.5 Exhaust Silencer Fires

Due to the triangular shape of the 'Deltic' engine, three exhaust manifolds were routed into a common drum before passing to the main silencer. The exhaust manifolds were cooled by a jacket through which engine coolant was circulated. The result of this was that unburnt lubricating oil passed from the cylinders tended to collect in the drum and silencer.

'Deltic' engines tended to pass relatively large amounts of lubricating oil by virtue of having exhaust ports in the cylinder walls and not having a particularly good oil scraper ring design. Over time this oil coagulated to form a tarry substance and there was a clear danger that this tar would ignite and potentially cause damage to the engine.

In normal service this oil burned off in small quantities when the exhaust gas temperature was high enough (hence the classic 'Deltic' blue exhaust). However, evidence showed that fires occurred most frequently on Monday mornings after the locomotives had idled all day on Sunday engineering trains; working hard first thing on early commuter turns caused accumulated oil from the exhaust silencer to catch fire. The usual remedy was to switch the traction motors out and run the engine at full power until the flames went out or alternately by ignoring the problem altogether and letting the oil burn off!

9.3.6 Failure of Auxiliary Gearbox and Associated Equipment

The failure of auxiliary equipment, components in the main supplied by specialist sub-contractors, proved to be a serious problem by the end of 1959. Failures, in the main due to the equipment being inadequate for its task, took place so frequently that spare parts were soon exhausted.

The provision of an auxiliary gearbox driven by a power take-off shaft from the gearbox at the free end of the 'Deltic' engine proved to be both troublesome and expensive. The power transmitted to the auxiliary gearbox provided onward drives to the air compressor, one traction motor blower and the roof mounted radiator fan. Soon after delivery of the locomotives, English Electric engineers noted that at engine idling speed the gear trains in the auxiliary gearbox were very noisy and that considerable 'backlash' was evident due to the combined action of the machines and shafts involved.

The auxiliary gearbox itself required constant attention and frequent replacement to keep the locomotives running, the situation being aggravated by the inadequacies of the driving and driven shafts coupled to it. Failure of the main cardan shaft between the engine free-end gearbox and the auxiliary gearbox began in November 1959, setting the stage for a spate of such failures. Worse still, failure of the main shaft frequently resulted in severe engine damage when the broken and flailing shaft smashed adjacent engine coolant pipework, interrupting the flow of coolant through the engine resulting in severe overheating and, in the worst-case scenario, multiple piston seizures.

Apart from the main cardan shaft issue, early troubles were also experienced with the clutch plate and drive shaft between the auxiliary gearbox and the air compressor. Initially the clutch plate problem was dealt with by using spare plates; however, to keep the fleet in service, weaker clutch springs were used to permit the clutch to slip at high torque to increase clutch life at the expense of clutch linings, thereby easing the acute spares situation.

Failure of the flexible drive couplings was dealt with by fitting temporary solid shafts to keep locomotives in service as a stop-gap measure until replacement flexible couplings became available from Metalastik; because the replacement solid shafts had neither clutch nor coupling, brass bolts designed for easy shearing were fitted in the flange mounting to the auxiliary gearbox to protect the gearbox and the engine in case of air compressor failure.

In August 1959 a 'Statement of Condition of the Air Compressor Drive, 1200hrs 07/08/59' was issued by BR, as follows:

Loco   Condition of Drive (Borg & Beck clutch and Metalastik coupling).

D5900   Clutch, as built. Coupling failed, solid coupling fitted.

D5901   Clutch plate failed; clutch and flywheel replaced by solid shaft manufactured by Napier, splined shaft retained. Brass, then steel, shear bolts fitted to protect the gearbox and engine in case of a compressor seizure. Coupling failed, solid coupling fitted.

D5902   Clutch, as built. Coupling failed, solid coupling fitted.

D5903   Clutch and coupling, as built.

D5904   Clutch and coupling, as built.

D5905   Clutch and coupling, as built.

D5906   Clutch, as built. coupling failed, solid coupling fitted. N.B. Two couplings had been fitted on this loco and both have failed.

D5907   Clutch and coupling, as built.

D5908   Clutch and coupling, as built.

D5909   Clutch, as built. Coupling failed, solid coupling fitted.

By the end-September 1959 the flexible couplings in nine of the ten locomotives had failed.

A proposal to fit an electrically-driven compressor to these locomotives was apparently made as early as November 1959 but no further detail has been found on the subject.

In January 1960 a new Rotoflex-Metalastik coupling of larger size was adopted. This unit was simpler with sufficient axial flexibility to eliminate the troubles with the splined hub of the clutch plate. Fitting of the re-designed flexible drive (and replacement of the traction motor blower shaft, where required) was completed during depot examination periods by April 1960, as follows:

Date	Loco. No.
25/02/60	D5903
29/02/60	D5905
09/03/60	D5900
10/03/60	D5902/6
28/03/60	D5901/8
12/04/60	D5904/7/9

The situation surrounding compressor drive failures improved as a consequence of these modifications but auxiliary gearbox and main cardan shaft failures between the engine and auxiliary gearbox continued to occur. These incidents resulted in frequent checks being made to ensure that the gearbox was topped up with oil, together with annual checks for 'backlash' and general condition.

Between November 1961 and May 1962 of the six 'Deltic' engine failures reported five involved main cardan shaft failures, four of these causing damage to coolant pipes. Various reasons for failure were cited but never formally agreed; EE suspected metal fatigue but the shaft manufacturers said the shaft had bowed due to being fouled by nearby equipment. Subsequent tests on cardan shafts revealed a tendency to 'whirl' at about two-thirds of the 'Deltic' engine maximum rpm rating.

By late 1962, the toll on the 'Deltic' engines and the severe impact on locomotive availability became intolerable, failures in some cases being at very low engine service hours. Unsurprisingly, given the significant damage being inflicted on the 'Deltic' engines, Napier strongly suggested that the locomotives be taken out of service until the cardan shaft problem had been properly addressed.

Troubles with the radiator cooling fan drive also resulted in engine coolant overheating. One such instance in March 1961 (together with other issues) attracting the wrath of H.J.H. Nethersole, General Manager (Traction Dept), London, in an internal memo he sent to P.J. Dalglish (Napier, Acton) and S.C. Lyon (RSDD, Preston) on 17 March 1961:

In reading over reports on service problems, etc., with the Type 2 'Deltic' locomotives certain aspects are somewhat disturbing.

Fan Drive Gearbox
A report from Mr. Bradley on the 1st March indicates a loose gear on the drive of the fan. As you will no doubt recall there was considerable trouble with the fan drive about a year ago, and it was thought the modifications to it had apparently cured the trouble. Is the case reported merely an isolated one, or is it an indication that we can expect further trouble with this drive after a longer period of time? If the previous modifications have merely extended the time before the drive will fail then I think we should very seriously consider the design of driving the fan electrically and remove the rather difficult mechanical drive altogether. I would like serious thought to be given by your design engineers to this matter immediately. We have promised to put the Type 2 locomotives into a first class reliable position, and the 'zero-life' build was supposed to be doing this.

Are we now up against further trouble which will shake the BTC confidence in us? Last July we stated that after the present modification to the locomotive it would be found to be absolutely reliable, and it was inferred at the time that if we had further major problems the BTC might decide to scrap the locomotives altogether and claim a penalty from us. This would do us no good and we cannot face such a situation arising. I hope, therefore, that all concerned appreciate the seriousness of the trouble.

I cannot stress too strongly, particularly for the benefit of Mr. Lyon's department, the need to deal with these outstanding matters on an urgent basis to clear up everything in respect of the Type 2 locomotives before June this year when the Railways are reconsidering the whole question of continued service of these locomotives. Unless the BTC have confidence that the locomotives are now reliable they are likely to dispense with them, and we as a company cannot face that issue.

Nethersole's memo perhaps overstated the fan drive issue in itself but it clearly illustrated English Electric's worries regarding the future of the 'Baby Deltics' in particular and the company's reputation in general. The memo generated a significantly greater focus on the spares inventory being held by the company in order minimise the impact of any failures (particularly in identified areas of vulnerability such as auxiliary gearboxes, shafts/coupling, etc.) and especially for long lead-time items.

9.4 Electrical Issues
Only a few electrical problems were experienced on the 'Baby Deltics'.

The main generators gave some trouble during the second half of 1959 because of blackening of the commutators which caused rapid brush wear. Cleaning and re-grinding of the commutators, plus trials with other types of brushes, were undertaken.

Some issues with the main generator armature bearings necessitated their re-design. Due to main generator armatures becoming out of balance it was necessary for EE to rebalance these during overhaul of the power units after removal from the locomotives. It was suggested that the issue was due to problems with the engine overspeed governors which, although designed to trip at 1680-1700 rpm, were sometimes failing to do so. EE tested the main generators when new to 2000rpm. Investigation showed that the armatures had gone out of balance due to the movement of the varnish in the end windings.

Traction motor troubles were few and flashovers very rare.

9.5 Other Issues

9.5.1 Excessive Noise

Excessive noise and noxious fumes affected both the drivers and passengers and ultimately precluded the 'Baby Deltics' from the restricted King's Cross-Moorgate tunnels except in emergency situations.

In an internal EE memo penned by H.J.H. Nethersole, General Manager (Traction Dept), London, to S.C. Lyon, (RSDD, Preston) (amongst others) dated 17 March 1961, mentioned earlier, reference was made to sound insulation tests being carried out on one 'Baby Deltic' locomotive with a 'patched up arrangement' but with no further work undertaken. Nethersole commented:

> I understand the Railways are still worrying about this problem and I think we should get on with the job and see what can be done and what it will cost.

Lyon batted it straight back to Nethersole in a memo dated 22 March 1961:

> At a meeting with the Railways at Doncaster on Monday last, we were again pressed by representatives of the Line Traffic Manager to state when we were going to fit insulating material to these locomotives. As you are probably aware, the drivers are shutting down their engines on every possible occasion to combat this unpleasantness and the constant stopping and starting is thought to be one of the main reasons for the battery troubles at present being experienced. We hope you will be able to let us have your proposals at an early date.

Based on archive material, not much happened until October 1961 when one set of Revertex sound insulation material was delivered to Stratford for installation by BR manpower supervised by EE, with a further two sets due to follow. The remaining seven sets were subject to further instructions. D5909 was duly modified but no further sets of Revertex material had been delivered to Stratford by mid-November, presumably awaiting the views of an acoustics expert as to whether any improvement had resulted on D5909. On 25 January 1962 it was reported by the BR(CM&EE) at Doncaster 'that the fitting of insulated material to locomotive D5909 has not had the desired effect'. By this time, four additional sets of Revertex material had arrived at Stratford but no further locomotives were fitted with the insulation material prior to them being taken out of traffic prior to the rehabilitation programme.

9.5.2 Fumes

No archive material has been found with respect to action taken to reduce fumes. Being fitted with two-stroke engines, presumably the heavy smoking was considered to be a fact of life! Locomotive rostering to preclude use on the Moorgate workings appears to have been considered sufficient.

9.6 Management and Staff Opinions

9.6.1 Senior Line Management Comment

In his book *The Trials and the Triumph* (2012), Tom Grieves provided some very interesting insights into the dieselisation of the Great Northern services in general, and the 'Baby Deltics' in particular. Grieves started his career on BR as a Premium Apprentice at Doncaster in 1949 and was heavily involved in the transition of GN suburban services from steam to diesel in the late 1950s and early 1960s. Grieves comments:

> [*It was a*] Courageous decision to dieselise such a high profile operation involving the mass transportation of the city and bowler hat brigade. The Clean Air Act following the deadly London smogs may have influenced the urgency. To say that Hornsey, followed by Finsbury Park, remained in the spotlight to the early 1960s is an understatement.

Grieves described the 'Baby Deltics' as 'the most maligned class we had to maintain and operate' and provided some reasons as to why the criticism heaped on the class was disproportionate compared with other Type 2s operating in the Great Northern King's Cross area:

> The cab layout…together with the novelty of the T9 'Deltic' engine, one quarter (*sic*) in engine capacity of the widely-publicised Blue Wonder ['DELTIC'], attracted those drivers with inherent rail enthusiasm. The Hitchin men were particularly enthusiastic, but who wouldn't be after being subjected to L1s [2-6-4T steam locomotives] on the relatively high-speed outer suburban services. Following commissioning they were immediately put onto these outer suburban and Cambridge

services. They made a lot of noise and had very good acceleration and if the rail condition allowed the full exploitation of the available adhesion they took off in meaningful manner. They had, however, an Achilles heel in that the compressor was mechanically driven through a drive shaft and clutch plate, the only such application within the new contracts. Their general reliability was far better than the other Type 2s but when the drive to the compressor failed it became a dead locomotive and could not limp home, so 'Baby Deltics' ran to time or not at all. The fact that they were on higher profile services and had the most operation over two-track sections (so that when something went wrong they stopped the job) raised the profile to such an extent that they became widely criticised.

The most striking incident was when D5908 failed with its train straddled across Cambridge Junction (Hitchin) effectively stopping the whole of the East Coast main line for over half an hour, a matter made worse because its position made rendering assistance all that more difficult. Without doubt it was a design fault by possibly too high a gearing as the engines were high speed at 1500rpm for most of the time, compared to 500-600rpm with conventional in-line vee engines. Clutches started to need replacing more frequently and in desperation a Jaguar XK120 clutch was fitted by joint depot and English Electric purchase.

This significantly extended the life but then occasionally shaft failures started to occur.

Irrespective of these failures, they performed well on the outer suburban and were popular with the crews. Hitchin men working the tightest schedules were well satisfied and any suggestion of providing a reserve B1 [4-6-0 steam locomotive] or L1 met determined opposition.

As time passed, engine block liner failures started to occur; these rarely caused line failures as the depot fitted an aeration tube to indicate air leakage and the locomotives were withdrawn for engine change as required.

9.6.2 Drivers' Comments (derived from '34B 34C 34D Hornsey/Hatfield/Hitchin Loco.' [Facebook Group])

Passenger

They were withdrawn just before I started at Hitchin but all the drivers there said the (*Cambridge*) Buffets were a bit heavy for them and they struggled to keep time.

With only 1,100hp they were on the limit but their acceleration was pretty good and once they got going they seemed to manage. They were even booked on the afternoon Cambridge Buffet Express which was extended to Ely.

I recall as an apprentice at 34G going on a few test runs on the Cambridge Buffet and it wasn't exactly exciting.

As a driver I have driven quite a few EE locos (Class 55/40/20), and as a secondman on the 'Baby Deltics'. The one thing they all have in common was when opening the power controller from a stand they were susceptible to overloading and seemed a slow process with accelerating to full power. With a lower hp 'Baby Deltic' on a passenger service you only had to have a couple of signal checks or TSR slacks and it was a struggle keeping time.

Freight (Cadwell traffic)

I remember when I was at Hatfield in the early 1960's and was booked spare on nights being sent to Hitchin to cover the Cadwell LPG [liquid petroleum gas] train which usually had a pair of 'Baby Deltics' in multiple, it was a very heavy train I recall.

Remember that job well, but only with a single 'Baby Deltic'; if it had been raining it was a hell of a job reversing it into the gas sidings at Cadwell.

General

Never failed on one in five years (*1961-66*), before and after rebuild.

I failed twice with them, once with up morning suburban working from Royston when a couple of pistons came through the crankcase at Woolmer Green, the second time was on the down Whitemoor between Ashwell and Royston when we had an oil pipe burst.

9.6.3 Maintenance Staff

Their small size left plenty of space for maintenance, making them popular with depot staff.

9.7 Out of Traffic and Into Storage

From Autumn 1961, as the 'Deltic' T9 *engines* suffered 'major' failures they were progressively laid aside pending a review of future options ranging from a further programme of T9 modifications to decisions relating to alternative engine types which could be deployed. 'Minor' repairs continued to be undertaken by Napier into May 1962.

As there were sixteen engines in the pool, it took some time for the spare engines to be used up and for the first locomotive to be taken out of traffic and so it was not until February 1962 before D5903 was retired following a major engine failure in the absence of an immediately available replacement.

Three more 'Baby Deltics' quickly followed (D5909 [March], D5902 [mid-April] and D5904 [early May]). The short-term availability of spare engines following 'minor' repairs facilitated a slowdown in further locomotives being removed from traffic until D5908 (August 1962); in the absence of repaired engines the next four locomotives were 'withdrawn' at roughly monthly intervals. D5905, the last one in traffic, finally succumbed in June 1963.

Details of the demise of the 'Baby Deltics' during 1962/63 were included in an article entitled '16th June 1963 - D5905 Stored Unserviceable' by Andy Wylie published in *Deltic Deadline*, Journal of the Deltic Preservation Society (June 1983). Salient details from this article, plus additional information from BR Diesel Casualty statements, are given below:

Loco. No.	Date out of traffic	Failure details
D5903	13/02/62	Loss of coolant water, piston seized. N.B. Loco previously released with fresh engine on 07/02/62.
		Failed on 17.40 King's Cross-Baldock.
D5909	07/03/62	'Leg out of bed' (piston through crankcase).
		Failed on 17.48 King's Cross-Welwyn Garden City
D5902	16/04/62	Cracked cylinder liners and con rod damage.
		Failed on 13.30 King's Cross-Baldock.
D5904	01/05/62	Broken con rods and fractured crankcase.
		Failed on 07.05 Royston-King's Cross.
D5908	23/08/62	Defective turbo-charger.
D5906	05/10/62	Cracked cylinder liner fuel injector hole.
D5907	13/11/62	Metal deposits found in 'chip trap' (indicative of advanced engine disintegration).
D5900	25/01/63	Cut out in traffic, suspected piston seizure.
		Failed on 08.20 King's Cross-Baldock.
D5901	28/01/63	Cracked crankcase joint.
D5905	13/06/63	Metal deposits found in 'chip trap'.

Note:
1. *Railway Observer* (June 1962) reported D5902 as failing at Letchworth on 1 May 1962 whilst working the 06.38 Hitchin-Baldock, subsequently being rescued by a Type 1. However, D5902 had previously catastrophically failed on 16 April 1962 and subsequently noted at Stratford on 29 April 1962; the RO report probably relates to D5901 (see Section 8).

BRITISH RAILWAYS

BR 33582/2

Eastern Region — Sheet 1

STATEMENT OF DIESEL CASUALTIES (MECHANICAL/ELECTRICAL)

Week Ended 17.2.1962 — Kings Cross District

Date	Unit No.	Type	Depot	Time	P. or F.	From	To	Delay	Nature of Casualty	Remarks
11.2.62	D9003	33/3	Fins Pk	10.0	P	Kings Cross	Edinburgh	8 min	Defective boiler	Incorrect combustion
12.2.62	D364	20/3	Haymkt	12/20	P	Kings Cross	Newcastle	Nil	Defective boiler	Fuel starvation
12.2.62	D238	20/3	Gatesh'd	11/50	P	Kings Cross	Newcastle	Nil	Defective boiler	Defective drip valve No 1
13.2.62	D5062	11/1	Fins Pk	6.39	P	Kings Cross	Hertford	32 min	Vacuum brake defect	Brake handle on wrong side
13.2.62	D5903	11/3	Fins Pk	5/40	P	Kings Cross	Baldock	51 min / Train can.	Loss of coolant water	A2 piston seized
13.2.62	D8240	8/5	Fins Pk	10.8	F	High Barnet	Ashburton Grove	Train can.	Oil leakage	B5 cyl head leaking oil, all other heads slight leakage
13.2.62	D5000	DRC	Lincoln	1/55	P	Lincoln	Grantham	11 min	Unable to start No 1 engine	
14.2.62	51265	DRC	Fins Pk	6.57	P	Hertford	Broad St	Train can.	Brake pistons sticking	
14.2.62	D5062	11/1	Fins Pk	2.46	F	New Barnet	New England	180 min	Cables back of exhauster on fire	Short circuit
14.2.62	51265	DRC	Fins Pk	8/9	P	Finsbury Pk	Hatfield	Train can.	Loss of air	Broken air pipe on No 1 engine compressor
14.2.62	51296	DRC	Fins Pk	3/3	P	Kings Cross	Hertford	11 min	Pistons sticking	
14.2.62	D5605	13/2	Fins Pk	6/2	P	Moorgate	Dunstable	Train can.	Flashover on main generator	Cause not apparent
14.2.62	D9008	33/3	Gatesh'd	8/5	P	Edinburgh	Kings Cross	12 min	Defective heater pipe	Normal wear

D5902-4/6-9 were *officially* stored unserviceable en bloc on 25/11/62 (w/e 01/12/62), although some of these locomotives had effectively been stored in an unserviceable condition for many months prior to this date. D5900/1 following on 03/02/63 (4we 23/02/63). By March 1963 photographs of a line of eight stored 'Baby Deltics' at Stratford depot began to appear in the railway press.

D5901 was reinstated to traffic during 4we 23/03/63 and according to Brian Webb (1982) it was sent to Vulcan Foundry arriving on 6 March 1963; the reinstatement was presumably a paper transaction to facilitate its movement to Newton-le-Willows where the locomotive was to be used as a test-bed locomotive for English Electric's new 12UT engine. D5901 was despatched from Stratford to Doncaster Works sometime between 3 and 17 February with onward movement to Newton-le-Willows sometime shortly after 3 March 1963. R.M. Tufnall (1979) lists the transfer date of D5901 to Vulcan Foundry as 28 April 1963, although, given the re-instatement date of 4we 23/03/63, the Webb date seems most likely. For further details regarding the 12UT project see Section 10.2.

BR(ER) Statement of Diesel Casualties (Mechanical/Electrical) (Sheet 1 of 2) for week ended 17 February 1962. Note entry for D5903 on 13 February, the first 'Baby Deltic' removed from traffic pending rehabilitation work.

The final weeks of D5905 in traffic proved to be quite eventful. In late May 1963 it suffered minor collision damage; however, rather than being stored it was repaired, being released back into traffic on 7 June. Some sources suggest that this work was undertaken at Stratford DRS but a Colour-Rail photograph of D5905 at Doncaster Works taken on 6 June 1963, with a very creditable 'touch-up and varnish' appearance, indicates otherwise (see page 84). However, on 13 June, 'metal deposits found in chip trap' sealed its fate and it was removed from traffic. It was stored unserviceable on 16/06/63 (4we13/07/63). Quite what happened to D5905 over the ensuing two months is not entirely clear.

On 15 June 1963, one day prior to official storage, D5905 was seen being towed northbound through Retford by D5069 and it was assumed at the time that it was en route to Vulcan Foundry, Newton-le-Willows. However, a photograph of D5905 at Doncaster Works (see page 125) taken by Grahame Wareham seems to refute this idea; the photograph shows the 'Baby Deltic' alongside D5644 in the Works Yard with a somewhat 'loose' date of '1963'. Doncaster Works visit reports for 9 June and 7 July 1963 show D5644 as located in the Works Yard and in the Erecting Shop respectively, although D5905 was not mentioned on either date. D5069 was also noted in Doncaster Works on 7 July. It is suspected, therefore, that Grahame Wareham's photograph was taken some time between 15 June and 7 July 1963.

It might be assumed that D5905 would subsequently move directly on to Vulcan Foundry for rehabilitation, but another photograph refutes this idea too. This photograph shows D5905 (with D5904 and another clearly stored 'Baby Deltic') at 30A Stratford (on the site of the old Jubilee shed) in 'July 1963' (see page 125). In addition to this, Steve Perkins recorded D5905 at Stratford on 7 July 1963.

It would appear that D5905 moved from Doncaster Works to Stratford, where all of the other 'Baby Deltics' were stored (barring D5901), to concentrate them all in one location pending a final decision on their future, a decision which was finally made in late July 1963.

An internal English Electric memorandum dated 11 September 1964 included the comment, 'You are probably aware that when these locomotives were "shopped" to Vulcan Works for rehabilitation, the engines of nine locomotives had been removed by BR depot and all the pipework 'dumped' in a pile in the engine room.' This undoubtedly refers to D5900/2-9, all ultimately being stored at Stratford, albeit D5905 only very briefly; these nine locomotives were subsequently moved to Vulcan Foundry for rehabilitation during late July/early August 1963.

D5900/2-9 were reinstated during 3we 03/08/63 (28/07/63), another paper transaction supporting their movement to Vulcan Foundry. D5900-9 were all subsequently stored unserviceable during 4we 21/12/63 (01/12/63) retaining this status until reinstatement after rehabilitation.

Pre-Rehabilitation Storage History (Sources: DLRCs, official BR Performance and Availability statistics and DPS *Deltic Deadline*):

Status changes	Locos Stored (Cumulative)	Loco. Nos.	Comments
7 locos stored 1we 01/12/62 (25/11/62)	7	D5902-4/6-9	Bulk transfer to store although most locos had been effectively stored much earlier.
2 locos stored 4we 23/02/63 (03/02/63)	9	D5900/1	
1 loco re-instated 4we 23/03/63	8	D5901?	1 loco officially shown as being in Works 4we 23/03/63 to 4we15/06/63 (inc) (assumed to be D5901 which was transferred to Vulcan Foundry on 06/03/63)
1 loco stored 4we 13/07/63 (16/06/63)	9	D5905	
9 locos re-instated 3we 03/08/63 (28/07/63)	0	D5900/2-9	
10 locos stored 3we 21/12/63 (01/12/63)	10	D5900-9	

9.8 Storage at Stratford

Above left: D5909/07/06/04/02/08/00/03, 30A Stratford, 6 March 1963. View of the line-up looking east with the No.2 Carriage Shop in the distance. (Nigel Petre)

Above right: D5902, 30A Stratford, 13 May 1962. (Colour-Rail)

Below left: D5902 and D5903, Stratford DRS, November 1962. (Author's Collection)

Below right: D5904 and D5909, 30A Stratford, 1962. (Rail-Online)

Above left: D5909, 30A Stratford, 16 March 1963. (Ray Simpson)

Above right: D5907, 30A Stratford, 16 March 1963. (Ray Simpson)

Below left: D5906, 30A Stratford, 16 March 1963. (Ray Simpson)

Below right: D5904, 30A Stratford, 16 March 1963. (Ray Simpson)

Performance and Service Problems (1959–63) • 121

Above left: **D5902, 30A Stratford, 16 March 1963.** (Ray Simpson)

Above right: **D5908, 30A Stratford, 16 March 1963.** Unique high level electrification warning flashes. (Ray Simpson)

Below left: **D5900, 30A Stratford, 16 March 1963.** (Ray Simpson)

Below right: **D5903, 30A Stratford, 16 March 1963.** Devoid of bogie dampers. (Ray Simpson)

D5909/07/06/04/02/ 08/00/03, 30A Stratford, 7 April 1963. Stratford shed had closed to steam in September 1962, but the top of the coaling stage can still be seen in the distance, above the third locomotive. The Diesel Repair Shop is visible in the left far distance. (Alec Swain [Transport Treasury])

D5903/00/08/02/04/ 06/07/09, 30A Stratford, 1963. Contrary to popular belief, there was never an occasion when all ten were lined up here together. Nine is a possibility either when D5901 arrived in January/ February 1963, or, when D5905 arrived in June 1963. (Rail-Online)

Performance and Service Problems (1959–63) • 123

Left: **D5908, D5902, D5906 and five others, 30A Stratford, June 1963.** Ongoing storage pending transfer to Newton-le-Willows but now re-positioned adjacent to the 'B' & 'C' Diesel Sheds. Looking in a south-westerly direction, and on the site of the former Jubilee steam shed demolished earlier that year (Steam Road No.6). (Author's Collection)

Below left: **D5906 and D5909, 30A Stratford, Undated.** Just north-east of the old Jubilee shed site and believed to be July 1963, with D5906/9 awaiting imminent movement to Newton-le-Willows. (Rail-Online)

Below right: **D5902 and D5908, 30A Stratford, 20 July 1963.** Ready for movement north with tail lamp attached. According to D5908's DLRC, it arrived at Vulcan Foundry three days later, presumably with D5902. (Nigel Petre)

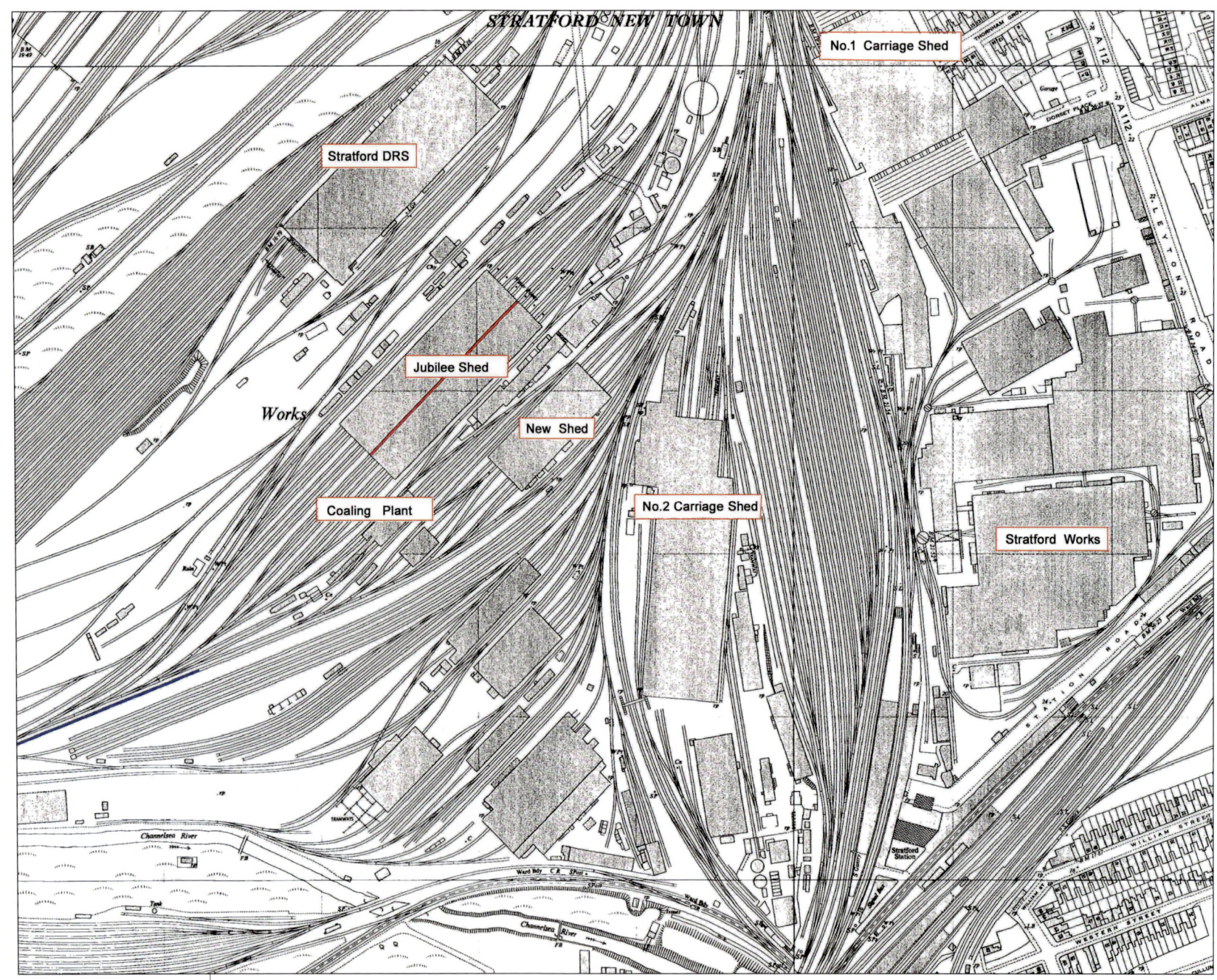

Stratford DRS and Depot. From Ordnance Survey 1:2500 map, composite 1953-66.

The line of eight 'Baby Deltics' shown in the previous photographs during the period March to May 1963 were stored on the line highlighted in blue towards the bottom left of the map. By June 1963 the 'Baby Deltics' had migrated to the site of the old Jubilee steam shed (Road 6, marked in red). The north-west half of the Jubilee shed (involving Roads 7-12) was demolished in 1959 to make way for the new Diesel Depot ('B' and 'C' sheds) which opened in August 1960. The south-east side of the Jubilee shed (Roads 1-6) was demolished in early-1963 after steam finished at Stratford in September 1962. In July 1963 Road 6 was used to store the 'Baby Deltics' where they were prepared for movement to Newton-le-Willows. Roads 1,2 & 6 were subsequently removed, leaving Steam Roads 3, 4 & 5 for stabling diesels. (© Crown Copyright and Land Information Group Ltd 2020 [Old-Maps.co.uk Ref.778604139]).

9.9 D5905's Perambulations

Above left: **A 'Baby Deltic' being hauled northbound past Marshmoor (south of Hatfield) by B1 4-6-0 61070, 1963.** Could this be D5905? There are only two known 'Baby Deltic' northbound 'dead-in-tow' movements through Hatfield in 1963. Firstly, D5905 in May moving to Doncaster Works for accident damage repairs (returning south in early June in spruced up condition and probably under its own power). D5905 moved north to Doncaster Works again in July 1963 following major engine failure; this movement is known to have involved D5069 itself destined for the Works for Classified Repair. The photograph is, therefore, believed to show New England's 61070 hauling a grubby D5905 in late-May 1963.

One other possibility is D5901, which has for long been assumed to have moved direct from Finsbury Park to Doncaster Works in February 1963 (for onward movement to Newton-le-Willows). However, sighting information provided by John Stretton positions D5901 at Stratford on 3 February 1963, which suggests that its transfer north was more likely via March rather than Hatfield. In any case the extensive tree foliage precludes this being a February shot.

D5905 was officially recorded as being in Doncaster Works from 29/05/63 to 07/06/63 so this is probably 29/05/63. (Jack Ray [courtesy Stuart Ray])

Above right: **D5905 and BR/Sulzer D5069, Retford, 15 June 1963.** En route to Doncaster Works. D5069 stayed at Doncaster Works until 25/07/63 receiving an Intermediate repair. (Brian Lee [via Rail Photo Archives])

Below left: **D5905 and Brush Type 2 D5644, Doncaster Works, 1963.** Same pattern of oil stains and blemishes as previous photograph. D5644 was in Works for its 2nd Intermediate repair (14/05/63-10/07/63). Photograph believed to be late-June 1963. (Grahame Wareham)

Below right: **D5905, D5904 and D590x (possibly D5903), 30A Stratford, July 1963.** On the site of the former Jubilee steam shed, three 'Baby Deltics' are being prepared for movement to Vulcan Foundry. (Colin Whitfield [Rail-Photoprints])

9.10 Transit to Vulcan Foundry

The October 1963 edition of the *Railway Observer* states that ' all ten [sic – nine] D59xx were moved from Stratford during the last two weeks in July. They were hauled to Doncaster and then proceeded to Vulcan Foundry early in August.'

Loco. No.	DLRC Information (VF Repair start date)	Comments
D5900	Vulcan Foundry: 30/07/63	Moved with D5907 (same VF date as D5907).
D5901	No DLRC record found	Arrived VF 06/03/63 (Brian Webb), hauled solo.
D5902	No DLRC record found	With D5908 (noted at Cambridge on 20/07/63).
D5903	Vulcan Foundry: 01/08/63	With D5904 (same VF date as D5904).
D5904	Vulcan Foundry: 01/08/63	With D5903 (same VF date as D5903).
D5905	No DLRC record found	Possibly with D5903/4?
D5906	Vulcan Foundry: 27/07/63	With D5909?
D5907	Vulcan Foundry: 30/07/63	With D5900 (same VF date as D5900).
D5908	Vulcan Foundry: 23/07/63	With D5902 (noted at Cambridge on 20/07/63).
D5909	No DLRC record found	With D5906 on 27/07/63?

Notes:
DLRC   Diesel Locomotive Record Card.
VF   Vulcan Foundry, Newton-le-Willows.

Several publications suggest the 'Baby Deltics' went to Doncaster Works after departure from Stratford but a total lack of sightings at the Plant (other than D5901) suggests otherwise. A short stop-over in the Doncaster area for crew/locomotive changing purposes is a possibility, of course, prior to onward movement to Vulcan Foundry.

Chapter 10
REHABILITATION

10.1 Options

During 1961/62 operational reliability and availability problems persisted. Whilst considerable effort was applied to resolve the issues, the ongoing impact of failures on locomotive availability resulted in more radical options being considered:

1. Substitution with an EE 1,250hp 8CSVT engine.
   See EE Drawing P.3200/xxx in *Deltics Super Profile*, R.M. Tufnell, 1985, p54).
   Modern Railways (July 1962) commented:

   > It is very strongly rumoured that English Electric is about to modify the whole class by substituting for the high-speed 'Deltic' engines a new charge-air-cooled version of the 1,000hp V8 engine used on the Type 1s. The new engines, of which ten are said to have already arrived at Stratford, are believed to be designated 8CSVT and to have a continuous rated output of between 1,100 and 1,250hp at 850rpm.

2. Substitution with a Napier 'Deltic' 1650hp D18-25 engine.
   See EE Drawing P.3200/403 in *Deltics Super Profile*, R.M. Tufnell, 1985, p53).
3. Deployment with the new 1550hp 12CSUT engine being developed by EE.
   See EE Drawing P.3200/428 in *Deltics Super Profile*, R.M. Tufnell, 1985, p27).
4. 'Back to basics' rehabilitation of the existing Napier 'Deltic' 1,100hp T9-29 engine.
   See EE Drawing P.3200/059 in *Deltics Super Profile*, R.M. Tufnell, 1985, p25).

Although the focus of attention of all these options was on the engine, it was ultimately recognised that the other problematical features would also have to be dealt with either by the substitution of alternative equipment or by rehabilitation, in particular the auxiliary gearbox and associated drives.

10.1.1 EE 8CSVT engine (1,250hp)
As early as July 1961 BR suggested that the 'Deltic' engines be replaced by an uprated version of the conventional EE 8SVT 8-cylinder Vee-type engine, as used in the 1,000hp Type 1 fleet (D8000 series).

Pro:
- The EE Type 1 fleet had proved to be the most consistently reliable locomotive in BR's diesel fleet.

Con:
- This option would add eight tons to the locomotive weight taking the gross weight of the locomotive above 80tons.

It was a subject that kept arising, no doubt due to the obvious extremes of the EE Type 1 and 2 reliability and availability statistics. Indeed, in June 1962, a fully developed 8SVT scheme surfaced offering either steam or electric train heating, together with a similar proposal using the uprated 8CSVT MkII engine.

10.1.2 'Deltic' D18-25 engine (1650hp)
During February 1962 English Electric put forward a proposal which involved the substitution of the turbo-charged T9-29 nine-cylinder engine with scavenge-blown 1650hp D18-25 18-cylinder unit as used in the recently introduced D9000-D9021 fleet (including the EE829 main generator, EE913 auxiliary generator and control system deployed on these locomotives).

Pros:
- A known commodity with BR (specifically at Finsbury Park) with respect to maintenance requirements.
- Commonality of spare parts and consumables (at Finsbury Park).
- Immediate avoidance of the turbo-charger reliability problems.

D5901 being hauled by D6790 through Patricroft en route from Doncaster Works to Vulcan Foundry during 1963. According to the Stevenage Locomotive Society *Aurora* Newsletter (March 1963) D5901 left Doncaster Works "on the 2nd or 3rd March for Vulcan Foundry"; Brian Webb (1982) records D5901 arriving at the Foundry on 6 March 1963. This photograph was very probably taken on the latter date. (J.R. 'Jim' Carter, courtesy of Chris Carter)

Cons:
- Excessive weight issue reduced compared to the 8SVT option, but still calculated to be 76tons overall.
- It would have placed the fleet in the Type 3 power bracket (where an English Electric product existed and was already in the process of rapid introduction).

10.1.3 Installation of a newly-developed 12CSUT engine (1550hp)

July 1962 saw work well advanced on a new 'revolutionary' EE diesel engine type known as the 'U' series engine. A prototype engine fitted with an EE835 main generator and EE913 auxiliary generator was already on the test bed at the EE diesel engine testing facility in Rugby. To fill a gap in the EE product range, the 'U' series engine was intended to provide EE with a robust lightweight high-speed diesel engine to supplement the RK/V heavyweight engine ranges and the 'Deltic' opposed-piston engines.

The company was keen to deploy one such engine in a locomotive for test purposes. With the 'Baby Deltics' being progressively taken out of service during 1962/63, English Electric sought permission from the BTC/BRB to use one as a test bed. English Electric produced detailed plans of the proposed re-engined locomotive, using the 12-cylinder unit set to deliver 1550hp and after some fairly protracted discussions the BTC eventually agreed to the use of a 'Baby Deltic' for test-bed purposes; D5901 was assigned to the project.

10.1.4 Further rehabilitation of the T9-29 engine (1,100hp)

The simplest option was to further modify and rehabilitate the existing engines to achieve reliability levels in excess of 87 per cent.

Although English Electric were promoting the 'up-graded' options, there was recognition that the existing 'Baby Deltic' locomotive was a likely BR preference, not least because of the extended time out of traffic expected with the more 'glamorous' re-engining options and the consequential protracted loss of Type 2 power in the King's Cross area.

Ultimately the plans to employ engines other than the T9-29 did not proceed, apart from the 12CSUT test-bed, and the T9-29 engine was retained albeit subject to major rehabilitation.

10.2 D5901/DP3

D5901 was nominated as the test-bed locomotive for English Electric's new metricated 12CSUT engine. All work was covered by EE Contract CCR1500. Readiness for traffic was scheduled for the end of September 1963.

The 'U' type engine was specifically designed for rail traction to provide a high-speed diesel engine of compact dimensions and moderate weight, and to have a high degree of reliability and durability. The power/weight ratio was a marked improvement over previous engines in the RK/V range and it was designed with a view to the development of rating levels up to 2,500hp. With respect to D5901, the original intention was to install the 12CSUT engine at 1800hp; however by February 1963 it had been decided by EE that, at least initially, it would be rated at 1550hp.

Brian Webb (1982) records that two of the 12CSUT engines were built although R.M. Tufnell (1979) indicates three. The leading particulars of the 'U' type engine were as follows:

Number of cylinders	12
Vee-form angle	45°
Cylinder bore/stroke	7¹¹⁄₁₆in x 8½in
Continuous traction rating to BSS 2953: 1958	1550hp at 1500rpm
Normal speed range	600/1500rpm
BMEP at traction rating	173lb/in²
Piston speed at 1500rpm	2125ft/min
Weight of engine	17,400lb
Weight per bhp	11.2lb/hp

Comparing the leading particulars between the existing 'Baby Deltics' and the re-engined locomotive revealed the following:

Detail	Existing 1100hp Loco	Proposed 1550hp Loco
Weight in working order	73t 17cwt	79t 0cwt
Dry Weight	69t 0cwt	74t 5cwt
Engine	'Deltic' T9-29	EE 12CSUT
Rating (traction)	1100hp at 1600rpm	1550hp at 1500rpm
Maximum Speed	75mph	90mph
Maximum tractive effort	46,200lb	45,000lb
Continuous tractive effort	31,800lb at 9mph	26,200lb at 16.9mph
Gear ratio	63:17	58:19

D5901 arrived at Vulcan Foundry on 6 March 1963. The engine to be fitted was promised for delivery from the EE Preston Works into Vulcan Foundry in July 1963, with the locomotive forecast to be ready for traffic by 30 September 1963.

On delivery the locomotive was due to be numbered DP3 to prevent any confusion with the 'Baby Deltic' fleet. DP3 would have been a substantially different locomotive capable of more demanding duties but with a more restricted route availability as a consequence of its axle loading of 19¾tons. Agreement in principle was reached between BR and EE for the locomotive to be used on London Liverpool Street-Kings Lynn services to make full use of its traction capabilities.

To enable the re-engining of D5901 a considerable amount of re-design work was undertaken, including:

- Underframe modifications to provide an enlarged engine well.
- Enlargement of the engine room floor area to accommodate the larger engine with its six mountings.
- Consequential repositioning of internal equipment (e.g. compressor, exhausters, traction motor blowers, pumps, radiator, etc. together with associated pipework, wiring and mountings).
- Superstructure alterations (including modified roof accessibility to cater for repositioned engine-room equipment, ventilation improvements (new air intakes at cantrail height and an additional air-filter louvres at No.2 end).
- Upgraded bogies (traction motor gear ratio change, and new primary and secondary springing, compensating beams and axleboxes) to suit the 20-ton axle-loading.
- Full updating of the control equipment.

D5901/ DP3 Side Elevation (with '1-2' side illustrated). Additional radiator grilles, additional air filters towards No.2 end and boiler air intake grille duplicated on '1-2' side. Position of engine exhaust port not known. (*Modified from* BR Main Line Locomotive Layout Diagrams, *Undated*)

No.1 End

No.2 End

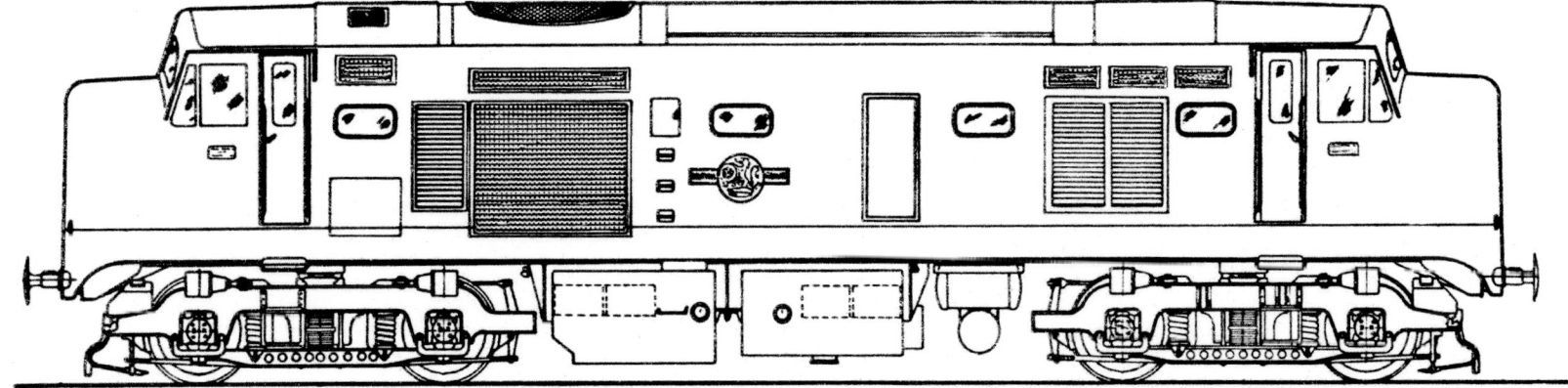

- Up-dating or replacement of selected items of major equipment (e.g. radiator);
- Cab soundproofing.

Work on the 'U' type engine ran into difficulties during May 1963 and its delivery to Vulcan Foundry was officially put back to early August. Further slippages meant that by early October 1963, D5901 was structurally complete but no further work was possible until receipt of the engine. The engine was assembled by early November, and, following testing, was scheduled for delivery by the end of the month.

Internal organisational changes within the Diesel Engine and Traction Divisions of English Electric during November 1963 resulted in the whole 'U' engine project being terminated, leaving D5901 stranded minus engine at Vulcan Foundry. Minute 63/73 of the BRB Technical Committee Meeting (31 December 1963) recorded:

> The Committee noted a letter (21/11/63) from the CME recording the English Electric Co's decision not to proceed with the development of the "U" high-speed engine and that the tenth Type 2 "Deltic" would be fitted with a standard rebuilt 9-cylinder engine.'

On 6 December the formal order was issued to convert D5901 back to its original form with a Napier 'Deltic' engine and Working No.CCR1500/53 was issued to cover the rehabilitation of the mechanical parts of D5901 to original condition.

Two days earlier, an EE order (CCR1500/40) had been issued by the Traction Sales and Contracts Department for the storage of a number of items of electrical equipment at Preston Works made 'redundant' as a result of the cancellation of the DP3 project. This document is particularly interesting in that it gives an insight into the electrical equipment intended for trialling and/or use in DP3, a project which tended to focus on the 12CSUT engine rather than the other critical components. Details of equipment destined for storage were:

1 EE835/5E main generator including spares.
1 EE913/1A auxiliary generator with overhung blower, including spares.
2 EE841A main generators.
2 EE913/2A auxiliary generators.

5 Gearwheels (58 teeth) with wheels/axles.
5 Pinions (19 teeth) with traction motors.
1 EE750/26G traction motor blower motor with Aerex blower.
1 Aerex 040 Blower with pulleys and two spare driving belts.

Some comments:

- The list included a totally new main generator Frame Number (EE841), a number not seen before or since.
- The EE835 main generator Frame Number indicated further mechanical and electrical variants, possibly including duplex lap winding.
- Three main generators and three auxiliary generators (two variants each) were included in the list for just one locomotive, maybe to ascertain the best engine/main generator/auxiliary generator combination, and/or, for equipment availability insurance purposes. The three sets of electrical equipment might support Tufnell's assertion that three 12CSUT engines were built to give three full power unit sets.
- The drive for the No.1 traction motor blower was evidently going to remain mechanical, albeit via pulleys and belts rather than via the infamous auxiliary gearbox.
- The traction motor Frame Numbers were not specified but it is believed the original traction motors were to be retained.

In many ways, the cancellation was a very unsatisfactory outcome; if the project had been pursued through to fruition it is arguable that a better engine might have been provided for the High Speed Train fleet, rather than the Paxman units actually specified.

10.3 Rehabilitation of D5900/2-9

10.3.1 Agreement to Rehabilitate

BTC/BRB documents dealing with the rehabilitation of the 'Baby Deltics' are, somewhat surprisingly, few and far between, certainly compared with similar information regarding the NBL diesel-electric Type 2s, the Metrovick Type 2s and the Brush A1A-A1A Type 2s. Entries in the BTC/BRB Works & Equipment and Supply Committee Minutes would have been expected.

The BRB Technical Committee Meeting dated 21 March 1963 (Minute 63/34 'Deltic Diesel-Electric Locomotives') was one Committee where reference to the 'Baby Deltic' rehabilitation was made, albeit briefly, as follows:

The Committee noted that because of unsatisfactory performance nine of the ten Type 2 locomotives required to have their engines rebuilt incorporating certain modifications. This will be done by the manufacturers free of cost. At the request of the manufacturers the remaining locomotive [D5901] will be fitted experimentally with a new high-speed engine (12 UT type) without cost to the Board.

It will be noted that this Minute only referred to the Napier engine rehabilitation and not general refurbishment. With no engine rehabilitation capital cost to BR, this would perhaps explain the minimal commentary in the BRB Committee papers, with general repair work beyond the engine being accounted for within the revenue account.

Withdrawal from traffic in 1962/63 was entirely the result of engine issues and it is probably not surprising that early thoughts on rehabilitation were limited to the Napier T9 engine. As is well known, however, the actual repairs undertaken on the 'Baby Deltics' amounted to a fully comprehensive Heavy General Repair (although it was never officially described as such); the reasons for the 'full-blown' upgrade was due to four factors:

- Although many issues affecting the 'Baby Deltics' were engine-related, there were other components requiring major attention to ensure reliable locomotives going forward (notably the radiators and auxiliary gearboxes and associated drives).
- There was a need to upgrade the locomotives to reflect modern working conditions and practices (sound insulation, deployment of route indicator boxes, etc.).
- Upgrades identified following general experience with diesel locomotives in traffic (e.g. 'Winterisation').
- Locomotives were stored at Stratford for extended periods (up to 18 months) and suffered from general deterioration and vandalism.

It appears that the original intention was for English Electric to refurbish and modify the first locomotive at Vulcan Foundry so that all the work required could be properly identified, specified, quality-controlled and tested. Subsequent locomotives would then be dealt with by BR, presumably at Doncaster Works, with reconditioned components supplied to BR by EE or the relevant secondary contractors.

By June 1963, however, EE had succeeded in persuading BR to allow them to undertake rehabilitation work on the whole class. This date is interesting in that it was mid-June when D5905 moved to Doncaster Works, perhaps with a view to engineers giving a 'Baby Deltic' the once over to determine the work involved once English Electric had completed the pilot repair. With EE ultimately acquiring the full contract, D5905 returned to Stratford pending 'squadron' movement of the class to Newton-le-Willow which occurred in late July/early August 1963.

With EE securing the contract work started on generating detailed work specifications.

10.3.2 Rehabilitation Work – Overview

All work undertaken as part of the rehabilitation process was covered by EE contract CCR1351.

When all of the work had been specified, the various activities were allocated specific Working Nos. Work undertaken on Electrical Equipment was grouped under Working Nos. CCR1351/21 to 29 and Mechanical Parts work under Working Nos. CCR1351/50 (for D5900) to CCR1351/59 (D5909), but obviously excluding CCR1351/51 (D5901).

Given the amount of work undertaken on the Napier engines Working Nos. in the range CCR1351/01 to CCR1351/19 were probably used for this work but this has not been confirmed. CCR1351/20 and CCR1351/33 are known to cover the corroded axle journal repairs on D5909 and Radiators respectively.

These work groupings will be referenced in the following sections.

10.3.3. Napier T9 Engines

Napier overhauled and rebuilt the 'Deltic' engines to comply with the firm's latest standards, the first engine arriving back at Vulcan Foundry in March 1964 for 200-250 hours testing in D5908 which ended up being the third locomotive delivered in the refurbishing programme.

No detailed specification has been found covering the full set of modifications carried out on the T9 engines, but it will undoubtedly have been an extensive list. It is known however, that new cylinder liners, crankshafts, pistons and connecting rods were manufactured and fitted. The liners were re-designed with tighter interference fits at each end so as to reduce flexing, and Mk.4A pistons were fitted with bolted-on crowns. The spare engines were also subjected to the same modifications, allowing the engine fleet to return to full strength.

10.3.4. Electrical Equipment

As with the Napier engine, no detailed specifications have been found covering the electrical equipment. The main and auxiliary generators, traction motors, exhauster motors, fuel pump motors, control equipment and master controllers were all sent to the English Electric, Preston Works, for reconditioning to as-new condition.

The Frame Numbers used for English Electric electrical equipment changed during their lifetime, notably the main generator (from EE835A to EE835D) but whether this reflected modifications undertaken as part of the rehabilitation process is unclear. The standard of engine-mounted electrical equipment was brought up to those applicable to the Type 5 'Deltics'.

10.3.5 Mechanical Parts

With regards to the Mechanical Parts aspect of the rehabilitation process, a letter was sent from the BRB CM&EE (J.F. Harrison) to the EE Traction Division, Bradford, on 19 August 1963; this is reproduced below in modified form:

Diesel Train Locomotives Type 2 'Deltic' 1,100hp - Re-engining and Refurbishing of Locomotives

With further reference to my letter of the 9th August and the telephone conversations between Mr. Shippen [Traction Sales & Contracts Dept., EE, Bradford] and Mr. Kilshaw [BR, Doncaster], I give below details of the work involved on nine of these locomotives and shall be glad to have your estimate in due course. On the question of the Stone's steam generators, I shall be writing to you under separate cover in the course of the next few days."

1. Lift body/underframe from bogies and place on stands.
2. Fuel tank. Remove from underframe; check and remove gauges and all inspection covers; clean internally and externally (including battery box); hydraulic test; paint; re-assemble in underframe.
3. Boiler water tank. Remove from underframe; remove and check gauges and all inspection covers; clean internally and externally (including battery box); hydraulic test; paint; re-assemble in underframe.
4. Clean underside of locomotive, including all air and vacuum pipes, conduits, etc. ready for inspection.
5. Test air and vacuum system.
6. Megger control, power and auxiliary cables.
7. Test control cubicle and master controller.
8. Remove roof.
9. Air filters. Remove all air filters; clean and re-oil; re-assemble in locomotive.
10. Radiators. Remove; clean internally and externally; test; re-assemble in locomotive.
11. Exhausters and motors. Remove from locomotive; return motor to EE Preston for overhaul; return exhauster to Reavell for overhaul and testing; re-assemble in locomotive.
12. Compressor. Remove compressor from locomotive; return to Westinghouse for overhaul; re-assemble in locomotive.
13. Fuel pump. Remove fuel pump motor; return to EE Preston for overhaul; re-assemble in locomotive.
14. AWS equipment. Remove and return to Chief Signal and Telecommunications Engineer for overhaul; re-assemble in locomotive.
15. Driver's air brake and vacuum brake valves. Remove for overhaul; re-assemble in locomotive.
16. Cooling water thermostat. Remove from cooling water system; test; re-assemble in locomotive.
17. Cooling water header tank. Remove tank; remove level switch; clean tank internally; re-assemble in locomotive.
18. Lubricating oil thermostat. Remove from lubricating oil system; test; re-assemble in locomotive.
19. Clean lubricating oil tank in situ.
20. Drawgear. Remove; clean and examine; re-assemble in locomotive.
21. Remove Stone's steam generator for attention as necessary.

22. Auxiliary drive gearbox, radiator fan, cardan shaft and traction motor blower fan. Remove for overhaul; re-assemble in locomotive.
23. CO_2 bottles. Remove; test pipework; check pull wires, etc.; fit re-charged bottles.
24. Speedometer generator. Remove speedometer generator, compensator gear, and both meters for overhaul; re-assemble in locomotive.
25. Insulate locomotive against noise.
26. Clean and repaint locomotive completely (interior and exterior).
27. Bogies. Remove traction motors and return to EE Preston for overhaul; remove wheels and axles from bogie; check axles for wear and test ultrasonically; clean and examine roller bearing axleboxes and, if satisfactory, re-fill with new grease; check tyres for wear; clean bogie frames ready for examination; gauge axle box guide liners; gauge bolster and transom wearing plates; check all rivets, pins and bolts for wear; check all springs; check brake pipework and equipment and test; fit new brake blocks; re-assemble bogie.
28. Replace body/underframe on bogies.
29. Final test.
30. Fit route indicators (single box as fitted to 'Deltic' Type 5); dispense with gangway.
31. Remove main and auxiliary generators from engine and return to EE Co. Preston for overhaul.

A number of comments are pertinent here:

- This specification was high-level and a significant amount of detail was left out (which must have made it difficult to assemble an accurate quotation).
- In the pure definition of the BR term 'Mechanical Parts', the list is consistent, with the exception of items 7, 11, 13, 27 (traction motors) and 31 which, beyond removal and refitting, were all electrical work (and subsequently dealt with as such in the allocated Working Nos). Any references to the Napier engine, main generator, auxiliary generator and traction motor blower work were correctly excluded, although for consistency the removal and refitting work should have been included.
- Reference to pipes, hoses and clipping, and electrical wiring and conduits were largely excluded presumably on the basis that the work was included within the Engine and Electrical categories).

English Electric responded to Harrison's letter on 4 September 1963 with a quotation (not 1962 as quoted in several magazine articles), but unfortunately this document has not been seen. However, a letter sent from the BRB CM&EE (J.F. Harrison again) to EE Traction Division, Bradford, dated 18 September more than adequately indicates progress:

> With reference to your quotation dated the 4th September, 1963, for the overhaul of nine of these locomotives in accordance with requirements set out in my letter of the 19th August, Items 1-29 and Item 31, my representative, Mr. Kilshaw, had a discussion with Mr. Scott [CM&EE, Eastern Region] on the 17th September to finalise the matter, and I understand that Mr. Scott has approved your quotation and will be writing to you direct in the course of the next few days.
>
> With regard to the route indicators, Item No.30 … I will be writing to you separately about this point.
>
> It is confirmed that as far as the tenth locomotive is concerned, i.e. D5901, which is being fitted with a 12 UT prototype engine, you will render a separate account for the overhaul of the locomotive which should be of a considerably lower figure than £4,278, bearing in mind that certain modifications and overhaul to control equipment had to be done to accommodate your prototype engine, and also that the actual work carried out would not be strictly in accordance with my letter of the 19th August.

The ER Sponsoring Engineer also sent a letter of acceptance of the quotation to English Electric on 18 September indicating that 'an order would shortly be issued covering this work' to Vulcan Foundry. No further mention has been found in the archives regarding the Stone train heating boiler 'question' mentioned in the 4 September communication.

By early October 1963, sufficient detail and costings (including the Napier engine and internal EE electrical equipment components) were available to allow English Electric to place internal orders on the various EE manufacturing Works.

Detailed below are the English Electric and Vulcan Foundry order details represented in composite form:

English Electric Order No.: CCR1351, dated 02/10/63

Vulcan Works Order Nos: 6396 to 6404 [for D5900/2-9 respectively]

Overhaul of 9 Type 2 locomotives.

Customer: BRB Eastern Region (Tender Date: 4th September 1963, BRB Order Date: 18th September 1963).

Delivery Required: Commence May 1964 at 2 per month.

Penalty for late delivery: Nil.

Maintenance Guarantee: Usual 12 months.

Summary of Equipment to be overhauled:

- 36 Existing EE533 Traction Motors (**Working No. 155CCR1351/21**),
- 9 Existing Control Equipments and Master Controllers (**Working No. 157CCR1351/22**),
- 9 Existing EE835 Main Generators (**Working No. 159CCR1351/25**),
- 9 Existing EE912 Auxiliary Generators (**Working No. 159CCR1351/26**),
- 18 Existing EE762 Exhauster Motors (**Working No. 159CCR1351/28**),
- 9 Existing Fuel Pump Motors (**Working No. 159CCR1351/29**),
- 9 Sets Mechanical Parts (**Working No. CCR1351/50/5900 and CCR1351/52/5902 to 59/5909**).

The Mechanical Parts Working Nos. covered the following:
1. Basic overhaul - Items 1-29 and Item 31 listed in BR's letter 19 August 1963. Price per loco: £3193.
2. Fitting of Route Indicators to BR Doncaster Design (Item 30). Price per loco: £306.
3. Replacement of missing parts and repairs to damaged items. Price to be submitted separately for each locomotive after joint examination by Vulcan and BR (ER) Resident Inspector.

10.3.6 Contract Additions

Inevitably as the work progressed various changes in the contract were made. A joint BR/EE Meeting was held at Vulcan Works on 10 February 1964 to discuss matters arising and additions to the contract agreed in September 1963. There was quite a substantial amount of 'scope creep' in the Mechanical Parts specification. The additional items were included under Working Nos. CCR1351/60 to CCF1351/74 (at least). Some of the changes (together with prices where known) are summarised below:

- Repair, overhaul and modification of cab door handles.
- Driving cab Therglas windows repaired or replaced.
- Driver's and Assistant's Seats re-trimmed and modified, including removal of the driver's right-hand arm rest (see Section 10.3.13). Seats were supplied by BR Doncaster.
- Driver's desk modification (including cutting away of the driver's desk to give additional knee room and repositioning of the horn valve handle) (Price per loco: £5.13. 0). **Working No.CCR1351/64**
- New type of driver's straight air brake valves (*Westinghouse Type F.A.1.*) (Price per loco: £96.12. 0). **Working No.CCR1351/60**
- Completion of drawgear modification (fitting of new/modified rubber rings and draw spring nuts) (already fitted to D5900/2/4/6/8/9) (Price per loco: £13. 1. 0.). **Working No.CCR1351/62**
- Protection Plates under buffer-beam (to prevent the screw coupling from damaging air pipes when allowed to swing down) (Price per loco: £6.17. 0). **Working No.CCR1351/63**
- Auxiliary gearbox modification, rehabilitation and relocation with associated drives (including gear replacement, bearing renewal, etc). **Working No.CCR1351/70**
- Mechanically-driven compressor replaced by electrically-driven compressor. **Working No.CCR1351/74**
- Radiator fan and clutch rehabilitation.
- Radiators cleaned internally and externally (with radiators found with leaking elements sent to Marston Excelsior for repairs).
- Exhaust lagging (partial replacement).
- Bodyside filters cleaned/re-oiled.
- Winterisation/Frost Protection (Price: £107). **Working No.CCR1351/61**, including:

Rehabilitation • 135

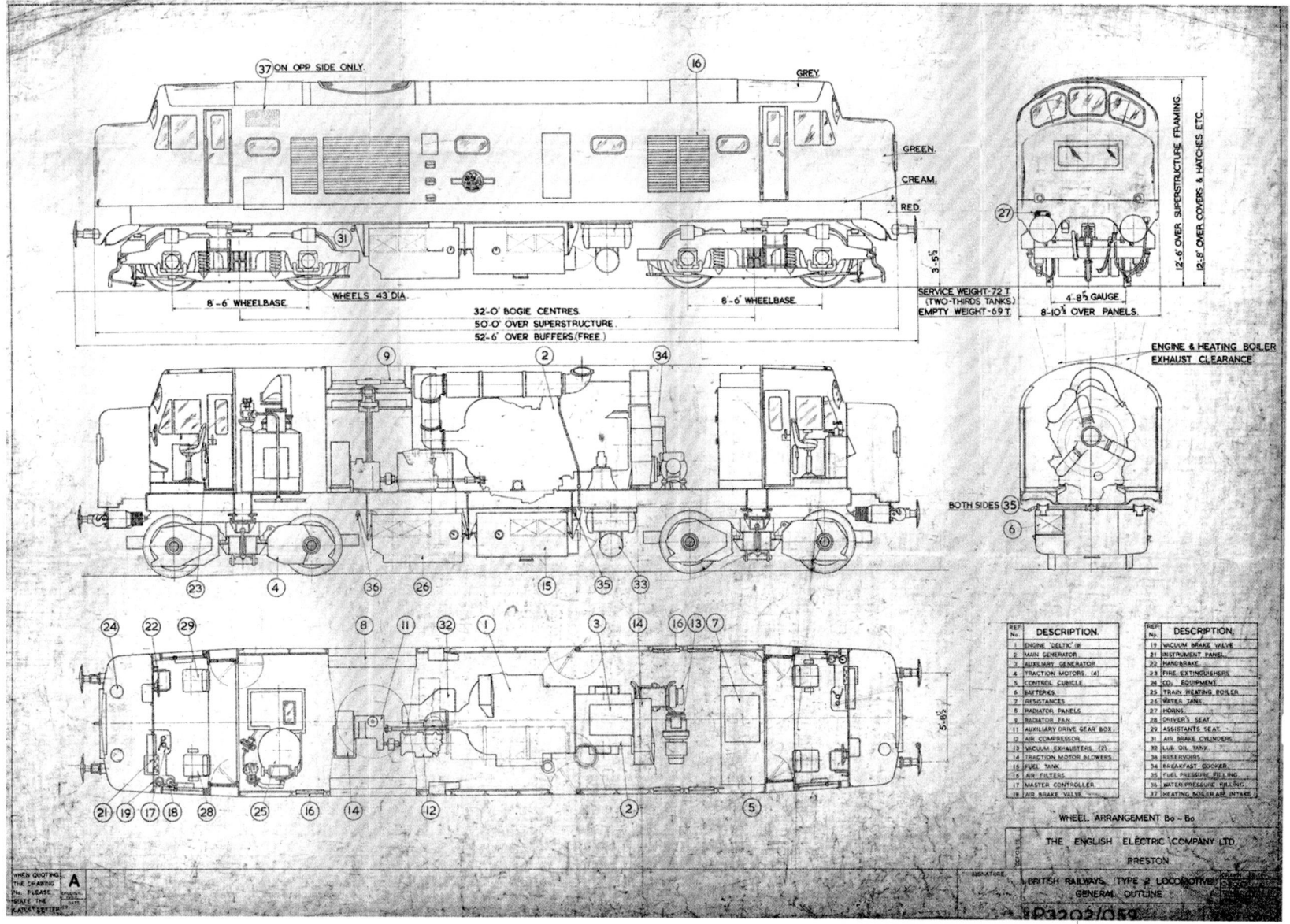

English Electric drawing illustrating side and top elevations. EE Drawing P.3202/059 (dated '19/4' presumably 19 April 1964). Interestingly whilst the drawing shows headcode boxes, the air compressor was still shown as mechanically driven and the livery was still the original as-built scheme, both being relatively late changes in the re-design process. (Grahame Wareham Collection)

- Lagging of boiler water feed pipes and valves.
- Re-routing of boiler feed water pipes (to minimise the length of pipework beneath the underframe).
- 'B'-side deflector plates (designed to prevent cold air entering through the air filters and freezing the pipes in the proximity of the steam generator).
- Train heating boiler modifications.
- Lagging of air system piping.
- Filling of anti-freezer in the suction pipe between the air strainer and the compressor.

- Fitting of automatic drain valves from the main air reservoir.
- Lagging of engine water cooling system piping.
- Bogies.
    - Axleboxes (external examination by Skefco).
    - Refitting of axlebox liners by welding to the horns (Price per loco: £92. 8. 0). **Working No.CCR1351/65.**
    - Bogie transom rubbing blocks. **Working No.CCR1351/71**
    - Armstrong shock absorbers (D5903 only, remainder already fitted).
- Refettling of trip-cock equipment.
- Steam-heating boiler (overhaul and modification by Stone).
- Boiler water tank filling connection modification (2½" fillers in place of 1¼") (Price per loco: £22.10. 0). **Working No.CCR1351/66**

A number of the modifications undertaken as part of the rehabilitation process are worth looking at in greater detail and are covered in the following sections.

10.3.7 Headcode Boxes

An internal EE (Traction Sales & Contracts Dept.) memo dated 7 August 1963 referenced a comment made by the BRB Mechanical Engineer (Design), Doncaster, as follows:

> It occurs to me that as these locomotives are now in your Works for overhaul and repair, it might be possible to incorporate the latest type of route indicators at each end of the locomotive … and to replace the marker discs at present fitted. Will you please let me have your comments on this point, so that I can make the necessary arrangements with regards to drawings and the supply of material.

Follow-up internal memos dated 12 August from the Rolling Stock Design Department to the Traction Sales & Contracts Dept. and the Chief Estimator indicated two possibilities:

> We have been requested… to examine the possible incorporation of route indicators to the above locomotives and propose to submit two schemes. 'Scheme A' having centrally mounted route indicators, as being fitted to the present Type 3 locomotives. 'Scheme B' having [two] separate route indicator boxes as was fitted on the early Type 4 locomotives [D325-44]. We would be obliged if you will submit prices in accordance with the following list of work involved:

Scheme A. Centrally mounted indicators.

- Remove the gangway corridor connections (one each end).
- Remove eight marker lights (from each end) complete with their associated cables and conduiting.
- Remove all marker light discs.
- Remove for repositioning the following: two access ladders, three CO_2 bottles (two at No.1 end and one at No.2 end), together with the CO_2 piping, clips, mounting brackets and operating gear.
- Cut away all gangway housing framing and panels; supply and fit three transverse angles, two vertical angles and fit and weld new panelling over apertures.
- Fit two route indicator boxes, one each end complete with new cables and conduits.
- Reposition the CO_2 bottles complete with operating gear, new piping, clips and mounting brackets.
- Repaint.

Scheme B. Independent route indicator boxes.

- Remove right marker lights (four each end) and the associated cables and conduiting.
- Remove all marker light discs (8 off) four each end.
- Remove access ladders.
- Remove for repositioning three CO_2 bottles (two on No.1 end and one in No.2 end) together with the CO_2 piping, clips, mounting brackets and operating gear.
- Cut away two vertical angles, four transverse angles and the associated panelling.
- Fit four route indicator boxes, two each end complete with new cables and conduits.
- Fit new handgrabs and steps in lieu of the access ladders for access to clean the cab windows and local steps for access to the route indicator winding gear.
- Reposition the CO_2 bottles complete with operating gear, screw piping, clips and mounting brackets.
- Repaint.

Drawings of the two options were submitted to the BR Sponsoring

Engineer on 6 September, and Scheme 'A' was subsequently selected (albeit without the access ladders refitted). English Electric initially quoted £450 per locomotive but this was subsequently reduced to £306 following an input from BR Doncaster regarding an alternative fitting method.

Further adjustments were made to the drawings prior to fitment of the indicator panels to ensure that the headcode blind selector and operating handles did not foul the fire bottles and to ease operation by BR staff. The route indicator equipment was supplied by Transport Engineering and Equipment Ltd, Lancaster.

10.3.8 Compressor

An internal EE memo composed by the Rolling Stock Engineer dated 23 March 1964 expressed serious concern about the mechanically driven compressor:

> Many of the earlier problems associated with this gearbox arose from the compressor take-off and flexible coupling. Whilst these latter items ran reasonably well in the later life of the locomotive there is little doubt that the torque fluctuations from the compressor adversely affects the life of the gearbox.
>
> This raises the whole question of the wisdom of taking the drive mechanically from a high-speed reduction gear box of light weight to a twin cylinder reciprocating compressor.
>
> Further the noise associated with this engine, particularly in stations such as King's Cross, when running only to create air pressure for the locomotive brake system, etc, has been a cause of complaint in the past.
>
> By changing to an electrically-driven compressor such as the 2EC38B, the first of which can be the surplus one from the 'U' engine conversion, and which is standard to the Type 3's, the gearbox problem and the long standing complaint re. the noise in stations can be eased.

An electrically driven compressor would require additional power take-off from the engine phasing gearbox to the auxiliary generator, but it was quickly ascertained that the auxiliary generator would be able to accommodate the additional load.

The 2EC38B compressors being delivered in advance for an order for Type 3 locomotives at Vulcan Foundry were identified for immediate use in the 'Baby Deltics' thereby avoiding any delays awaiting delivery of new equipment.

The cost of the compressors was about £570 each with motor, plus installation costs. Given the cost involved the approval of the General Manager, Traction Division, was required. In an internal memo dated 8 April 1964 the General Manager, Traction Division, commented:

> I had an opportunity of discussing this matter with Mr. Howard, Mr. Dowling and Mr. Collingwood at Vulcan on the 2nd and it was agreed that in spite of the extra cost it was the right policy and I confirm, therefore, having advised you on the 7th instant to go ahead with the proposed replacements.

BR were informed of the proposal on 6 April and it was duly accepted. Instructions were given to Westinghouse not to proceed with the recommissioning of the mechanically driven 2EC72 compressors, although in reality the instructions arrived too late.

In terms of costs, the following letter from the BRB (J.F. Harrison/ T.B. Maddison) to English Electric dated 15 June 1964 clarified the various issues:

> In your letter dated 6th April 1964 you requested permission to remove the mechanically-driven compressors on these locomotives, and replace them by electrically-driven ones, and I accepted on the understanding that there would be no charge to British Railways, and because I have always been in favour of fitting electrically-driven compressors to these locomotives.
>
> At a recent meeting held in Vulcan Works, you requested disposal instructions for the ten mechanically driven compressors, as, in your opinion, they were British Railways property; but Mr. Kilshaw (*BR*) advised that as you had replaced them free of charge, there was no point in returning this equipment to British Railways, apart from any other factors, no other locomotives are fitted with mechanically-driven compressors.
>
> Regarding your request for the cost of overhauling these compressors, amounting to £40 each, to be included along with other costs of repairs, my

answer to this is that the cost cannot be accepted, and you should offset these charges by the fact that you are receiving ten mechanically-driven compressors for your disposal.

May I make the point also that if you had requested permission for a design change at an earlier date than April 1964, there would have been no need to have had the mechanically-driven compressors repaired.

10.3.9 Auxiliary Gearbox

As part of the rehabilitation, reconditioning of the auxiliary gearboxes was essential, together with the Varley oil pump. A detailed inspection was undertaken during early 1964 and a list made of all defective parts. The inspection process identified many items requiring renewal, particularly severely worn gear teeth and the splines on the gear shafts. It was also noted that the original design using long cardan shafts and a large number of gear meshes, combined with the 'backlash' impact from of the several driven auxiliaries (radiator fan, a traction motor blower and an air compressor with very severe torque fluctuations) was the ultimate cause of the considerable damage found.

Each gearbox was marked to its own locomotive and, after stripping, the gearbox parts were cleaned, protected and stored separately, until required for re-assembly.

BR promised to make their stock of gearbox spares available to EE to assist with re-conditioning the boxes. Outstanding parts requirements were met either by re-conditioned stock items or newly manufactured replacements.

An internal EE memo dated 20 February 1964 succinctly described the gearbox problem:

To summarise, the design of the box is criticised in that there are too many splines, and gears mounted in this way cannot be expected to operate very satisfactorily. Neither has it proved very satisfactory to use such long cardan shafts connected by a train of gears, which must add up to excessive 'wind-up' and resultant hammering. It would seem advisable to consider individual load drives for the compressor, fan and blower, either by electric motors or hydraulic motors. These could be fed from the existing power output coupling on the engine.

By April 1964 the question of gearbox modifications following the introduction of motor-driven compressors was being discussed, specifically the removal of some idling gears and the compressor output shaft and gears with suitable blanking flanges fitted in lieu.

After re-assembly, each gearbox was coupled to an electric motor and tested under light load, to ensure that the lubrication functioned correctly and that there was no overheating of seals and bearings.

Following the re-introduction of the rehabilitated 'Baby Deltics' one complete reconditioned gearbox spare was held in stock.

10.3.10 Auxiliary Equipment Drive Arrangements

Originally the 'Baby Deltics' were fitted with Hardy Spicer cardan shafts between the engine and auxiliary gearbox but in 1964, as part of the rehabilitation process at Vulcan Works, it was decided to change to the Layrub shaft design in order to avoid any further instances of breakages and, perhaps more importantly, to avoid any collateral damage resulting from flailing shafts causing damage to pipework and associated loss of coolant and, on occasions, severe engine damage.

The new shaft fitted was a tubular splined shaft with Layrub 'Two-Four' series couplings at each end. Tests were carried out at Vulcan Works and it was accepted that these shafts should be fitted during rehabilitation but that a more substantial shaft with 'Six-Six' series couplings would be fitted at a later date when the required parts were available.

The 'Six-Six' coupling was a stiffer coupling and considered to be more suitable than the 'Two-Four'. The 'Six-Six' shaft was 17 per cent larger in diameter and 35 per cent thicker in section than the locomotive's original shaft, and it was dynamically balanced. Shaft vibrations due to secondary whirl, which caused the fatigue and failure of the original shafts, occurred on shafts within the 'Deltic' engine running range, but were eliminated on the Layrub equipment as the critical frequency causing whirl of the new shaft was above the maximum overspeed of the 'Deltic' engine.

The interim Layrub shafts with 'Two-Four' couplings were sanctioned, despite the additional costs, to avoid any further risk of failure of the original shafts.

Twelve interim sets of shafts with 'Two-Four' couplings were ordered, with a further twelve sets of shafts with 'Six-Six' couplings ordered for fitment later; one set of the latter equipment was supplied in advance for deflection testing purposes. Delivery was promised for end of November/early December 1964. New power take-off and gearbox flanges were manufactured to accommodate the new couplings.

Arrangements were made for the new 'Six-Six' equipment to be fitted to locomotives still at Vulcan Foundry at the time, with the remainder fitted at Finsbury Park by EE/Napier engineers during scheduled examinations. Strict alignment checking was stipulated following installation. Drive shaft guards required modification to accommodate the new equipment and full proximity testing was imposed to avoid any interference with other equipment.

From internal EE correspondence it appears that D5907 was the first locomotive fitted with a shaft with the 'Six-Six' couplings (at Vulcan Foundry), suggesting that D5901 was also fitted there, with D5900/2-6/8/9 retro-fitted at Finsbury Park from March 1965.

In practice, shafts with the 'Two-Four' couplings proved satisfactory in service and there were no breakages up to the point when the shafts and couplings were upgraded.

10.3.11 Coolant Flow Switch

As a 'belt and braces' action, a coolant flow switch was retro-fitted when the locomotives were rehabilitated. Although work to address cardan shaft breakages was given a very high priority, together with the provision of additional shaft guards to protect adjacent coolant pipes, BR and EE wanted to be absolutely sure that in the event of any future shaft breakages that no collateral damage would be sustained by the engine.

Coolant pipes, once fractured, could drain the coolant out of the engine at a substantial rate and, given that the engine would only shut down once sufficient coolant had drained out of the header tank, significant engine damage could already have occurred.

The new coolant flow switch monitored the rate of flow of coolant and acted to shut down the engine as soon as a flow rate interruption was detected, i.e. immediately following a pipe fracture.

10.3.12 Radiator

Despite the issues associated with the Marston radiators prior to rehabilitation, there does not appear to have been the same level of concerted effort applied to ensuring the reliability of the radiators following rehabilitation as directed towards most of the other items of equipment deployed in the 'Baby Deltics'.

Replacement of the Marston radiator elements by Spiral Tubes elements, as used in the English Electric Types 1, 3 and 4 (but not Type 5) fleets would have been a sensible option given their trouble-free history on these locomotives. However, patch-up and make-do with the Marston radiators seems to have been the order of the day for the 'Baby Deltics'.

In June 1964, only days before the re-entry into traffic of the first rehabilitated locomotive, the position was as follows:

- Seven radiator panels had received or were receiving new matrices, treated with '70/30 arsenical/brass' corrosion protection, at Marston Excelsior (at a cost of £440 per pair).
- To avoid continued corrosion trouble, the remaining seventeen (thirteen, plus the four spares at Doncaster) were to be either metal sprayed or dip-treated with Epoxy resin by English Electric (Accrington), following 'deep cleaning' (involving removal of the oil and water radiator blocks from the radiator frames, degreasing by steam cleaning, wire-brushing and hand-cleaning, if necessary).

An additional modification involving a ⅛in thick packing strip sweated onto the tube header plate was added later to help reduce further joint failures.

An internal EE memo dated 26 June 1964 commented that 're-tubed radiators have been used in the first two locomotives completed [D5904/9], and hence there will be no need to withdraw them from service for a radiator change'. However, another memo dated dated 29 June indicated that 'we had obtained the Railways permission to allow the second locomotive No. D5909 to be delivered on July 1st with a set of the original radiators fitted and that re-tubed radiators would be fitted to the third out-shopped locomotive, D5908.'

The 29 June letter also commented that permission had been granted for an eighth radiator to be fully re-tubed by Marston. Radiator Nos. 3/4/13-7/21 were fully re-tubed at BR's expense and further radiators may have been re-tubed at a later date following detailed inspection of their condition (possibly Nos. 1/2/23/4).

In the event the 'Baby Deltics' re-entered traffic with a mix of original (albeit cleaned internally and externally), treated or fully re-tubed radiators, a surprisingly haphazard state of affairs. The adverse effect of this less than satisfactory situation was rapidly realised as will be discussed in Section 12.2.2.

10.3.13 Sound Insulation

The subject of improved sound insulation on the 'Baby Deltics' arose before rehabilitation and it is believed that only one locomotive, D5909, was fitted with Revertex insulation material in late 1961 with nine further sets of material manufactured but not fitted. D5909 was fitted with insulation materials in the cab bodysides and cab bulkheads (including all access doors to the engine room and boiler compartment). The result of this work, however, was not deemed adequate by BR footplate staff.

On 1 August 1963 J.F. Harrison insisted that noise levels on the 'Baby Deltics' be given special consideration and instructed that sound insulation should be provided on all locomotives as part of the rehabilitation process.

Part of the D5901 conversion project, which was already underway by this time, included the fitting of screens behind the Driver and Assistant's seats. The seats had to be modified to accommodate the screens; with the screens fitted the seats were unable to rotate and to allow drivers and secondman access to the seats the inner arm rests had to be removed. In addition, the interior of the cab side panels and cab roof was sprayed with 'Limpet' asbestos ¾in thick and covered with glass cloth.

At this point the archive trail goes cold and whether the locomotives were indeed fitted with sound insulation materials is not known; the solution may have been the fitting of both Revertex and the D5901-related materials (asbestos and glass cloth). When D5901 was converted back to standard, no mention was made regarding removal of the asbestos and glass cloth from the interior cab roof panels.

Numerous photographs in this book show that draught screens were fitted behind all cab seats on all locomotives, necessitating the use of modified seats.

10.3.14 Repainting

The livery chosen for the refurbished locomotives conformed with the Type 5 'Deltic' locomotives, i.e. two-tone green with nose-end yellow warning panels.

10.4 Conversion of D5901 to Standard

At the point of cancellation of the DP3 project, work was already underway at Vulcan Foundry on the thorough rehabilitation programme of the remaining nine examples of the class. D5901 was effectively added to this contract and was converted, so far as was practical, back to an orthodox 'Baby Deltic'.

In order to avoid some of the cost English Electric looked to BR for concessions on some of the work required to effect an exact re-conversion. In terms of the superstructure, the concessionary areas to be pursued were identified in an internal EE memo between the Rolling Stock Design and the Traction & Sales Contracts Departments dated 23 December 1963:

BRB Type 2 Locomotives (D5901) – Re-conversion
We refer to your memo of 13th instant and set out in detail the parts we would like to retain to keep down the conversion costs…

(a) *Underframe Well*
This was extended at both ends to allow for the longer and lower engine and we feel that to convert back would be an expensive and patchy job. The front section will be suitably reinforced by folded sections to take one foot of the compressor and one mounting pad of the lubricating oil tank, together with suitable supports for the engine to gearbox drive shaft guards. The rear section will require a small modification to take the air entry of the No.2 end (engine-mounted) traction motor blower.

(b) *Interior Finish*
On 'A' side, the first two bays forward of the side service door have been dished in to allow for the mounting of the lubricating oil filter and strainer. To convert would require new panels of aluminium faced 'Onazote', removal of the brackets and supports and new supports added.

(c) *Radiator Shutter operating gear 'B' side*

The position of the screwbox has been moved through 90° so that the box is now mounted in the same place as the side structure, the operation is not affected in any way.

(d) *Fire Detectors*
These have been moved from the centre of the roof to the side caves and are mounted adjacent to the upper cant rail with the phials pointing in an upward direction.

(e) *Additional Air Filters*
Six filter panels 20inx10in have been added. These are mounted in the caves (3 each side) above the existing filter panels at No.2 end. To remove these would require the six louvre panels and the six filter housings to be cut away and new panels fitted and welded into the aperture.

(f) *Arising out of (e) (Lighting)*
Two of the interior lamps, one each side are repositioned to clear the filters. To position these together with the fire detectors (d) would require all the lighting conduits and cabling between the bulkheads and crossing the structure to be removed and replaced with new conduits and cables.

We trust the above will enable you to put over to Mr. N. Kilshaw, BR Doncaster, a case for retaining those parts of the locomotive as modified.

These topics were discussed by BR and EE at a liaison meeting held at Vulcan Works on 10 February 1964, with D5901 available for inspection. The decisions made regarding the six items identified for potential concessions were as follows:

BRB Type 2 Locomotives – Locomotive D5901
Points raised in connection with memorandum of 23rd December 1963 from RSDD and TS&C Bradford:

a) Underframe Well:
Mr. Kilshaw agreed with our personal [sic] that the well shall remain as converted and is satisfied with the method of mounting the compressor, lubricating oil tank and cooker by introducing folded section supporting members. Modifications required for new ducting also agreed on. Further investigation will be made to fit additional covering plates where personnel walk, if required.
b) Interior Finish: At the initial discussion it was requested that this should be converted back but on inspection it was agreed that the work involved did not justify this and therefore was acceptable.
c) Radiator Shutter Operating Gear (B side): It was agreed the revised position is fully acceptable.
d) Fire Detectors: The new position of these was agreed to.
e) Additional Air Filters: All these i.e. six filter panels adjacent to the generator and three above the radiator A side, are to be removed and the apertures filled in.
f) Lighting (arising out of (e)): It was agreed that the two interior lamps, one each side of the Generator section of the structure, be left in the position to which they have been moved due to inclusion of the filters.

Minutes from a further BR/EE liaison meeting held on 14 April 1964 included the comment:

BRB Type 2 Locomotives – Reconversion of Locomotive D5901
(b) Interior Finish: This will now be rehabilitated to as original as requested since the last meeting in BR letter of the 2nd March 1964.

With all the key decisions made, the English Electric Rolling Stock Design Department was able to specify the re-conversion work required to Vulcan Foundry enabling work to commence. Relevant details are given below and are based on an amalgam of internal EE memos circulated during April 1964:

Locomotive Superstructure
(a) Radiator compartment. The radiator [for the 12 UT engine] (together with mountings, partition and framing, compartment drains, water header tank, radiator and header tank piping), gearbox and traction motor blower mountings are to be

removed and replaced with the standard Type 2 components.

(b) Underframe and traction motor ducting. Apertures in the underframe left by the removal of the traction motor air ducting and generator air outlet duct, are to be made up and new pieces of ducting manufactured.

Remove [12 UT] engine and generator mountings and manufacture and fit new mountings.

(c) Superstructure. The roof, fixed section, between the radiator roof and engine roof is to be made back to the original drawing. The original removable roofs are to be refitted (complete with new seals).

The air intake louvres in the roof eaves on both sides are to be removed and the apertures blanked off.

Interior finish 'A' side, which was modified for the inclusion of the lubricating oil filter and strainer mountings, to be converted as original.

(d) Air/Vacuum Piping & Exhausters. The air piping is to be replaced in accordance with original drawings, except that new piping is required for the motor-driven air compressor. Fit and assemble the existing exhausters with new mountings.

(e) Coolant, Lubricating Oil and Fuel Oil Piping.

Refit original water header tank, lubricating oil tank and lubricating oil vent tank.

Reposition semi rotary water pump in the engine department.

Refit all piping between engine, radiators, tanks, etc (using new rubber flexible pipe connections).

Locomotive Rehabilitation
In addition to the work necessary in reconverting to the original design, this locomotive is to be rehabilitated in accordance with the schedule of standard work, items 1-24, 26-31 (BR letter dated 19 August 1963).

All the latest BR requirements are also to be carried out in accordance with instructions and drawings already issued for the other nine locomotives:

1. Sound insulation.
2. Modify driver's desk to give additional knee room and re-position horn handles.
3. Westinghouse F.A.1. type driver's brake valve, proportional valve and reducing valve to be reset.
4. Cab screens.
5. Protection plates (buffer beam).
6. Drawgear modifications (Spencer Moulton ring and special nut).
7. Water filling connection to water tanks 2½" and delete existing filling connection, blanking off at the tank.
8. 'Winterisation' Requirements.
9. Route indicators.

The bogies of D5901 were fully re-converted to become the same as the bogies on the other nine locomotives (including all modifications agreed with English Electric as part of the rehabilitation process) and involved:

- General restoration to original condition, with new wheel sets, tyres, axles, gear wheels (63 teeth) and traction motor pinion wheels (17 teeth).
- Original axle boxes, primary/secondary springs and equalising beams to be refitted.
- Axle box guide liners to be welded in position.
- Revised method for securing bogie transom rubbing blocks.

10.5 Rehabilitation and Associated Costs

It is unclear within the available archive material as to how exactly the cost of the rehabilitation work

D5901 undergoing rehabilitation at Vulcan Foundry, Newton-le-Willows. Standard roof sections and no residual evidence of additional grilles, so works seems to be well advanced in converting D5901 back to a standard 'Baby Deltic'. 1964/65 photograph.
(Colin Marsden Collection)

undertaken on D5900/2-9 was apportioned.

The Minutes of the BTC/BRB Works& Equipment Committee, Technical and Supply Committees provide no insights at all regarding the costs associated with each of the various rehabilitation options, not even the option finally selected, which is most unusual given that these documents are both readily available and seemingly comprehensive. At face value this might suggest that there was no cost involved which required authorisation by the BTC (which is clearly not the case), or, given that the upgrade could broadly be regarded as nothing more than a major 'General' repair of existing equipment (i.e. not involving any capital cost for re-engining as required with the NBL and Brush Type 2s), then presumably any costs could be 'hidden away' under revenue rather than capital costs.

It is generally known that all costs relating to the Napier 'Deltic' T9 engines were underwritten by English Electric to resolve long-standing engine (and turbo-charger?) issues. Correspondence indicates BR's acceptance of an English Electric quotation of costs covering the additional work required to put the 'Baby Deltics' back into first-class condition with respect to 'Mechanical Parts'.

What is not at all clear is the apportionment of the cost of work undertaken on problematical peripheral engine equipment (e.g. auxiliary gearboxes and associated drive). Logically I would expect that these costs to have been covered by BR.

Certain additional costs will definitely have been paid by the BR, including:

- Replacement of consumable or cannibalised parts not present on the locomotives at the time of their arrival at Vulcan Foundry.
- Any repairs to damage (e.g. the bodyside damage suffered by D5906 ['2-1' side]).
- Costs associated with the major repairs of the locomotives generally and components specifically which would have been due around this time (and repairs to auxiliary mechanical and electrical equipment would ordinarily have fallen into this category).
- Additional BR-specified up-grades including fitment of four-character headcode boxes, removal of gangway doors, safety up-grades, 'Winterisation' initiatives, etc.

10.6 Re-Delivery

Delivery details were presented by EE representatives at a BR/EE Liaison Meeting on 9 June 1964; the programme, taking into account EE works holiday shut down periods, was agreed and is reproduced below:

Re-Delivery.

Loco No.	Engine No.	Forecast T9-29 Engine Delivery Date to VF	Forecast Loco. Delivery Date to BR
D5904	1	Delivered	19/06/64
D5909	2	Delivered	11/07/64
Remaining Loco. Nos. not known at this stage	3	23/06/64	14/08/64
	4	30/06/64	21/08/64
	5	Mid-07/64	04/09/64
	6	End-07/64	18/09/64
	7	Mid-08/64	02/10/64
	8	09/64	23/10/64
	9	09/64	06/11/64
	10	10/64	27/11/64

With the exception of the final two locomotives this programme was broadly adhered to. D5904 and D5909 were both despatched to Doncaster Works for acceptance trials in July 1964; the final two, D5907 and D5901, were delivered in March and April 1965 respectively. The delivery of D5901 was delayed due to the time needed to undo the extensive modifications made to allow it to accept the 12CSUT engine.

10.7 Liquidated Damages?

Time out of traffic awaiting/undergoing rehabilitation:

Liquidated Damages.

Loco No.	Dates stopped pre-rehabilitation (*Deltic Deadline*)	Dates re-introduced to traffic (B. Webb)	Time pending/ undergoing re-habilitation (days)	Length of time out of traffic (ranking)
D5900	25/01/63	02/10/64	616	9
D5901	28/01/63	29/04/65	822	5
D5902	16/04/62	27/11/64	956	1
D5903	14/02/62	04/09/64	933	2
D5904	01/05/62	01/07/64	792	6
D5905	13/06/63	20/12/64	556	10
D5906	05/10/62	29/10/64	755	7
D5907	13/11/62	31/03/65	869	3
D5908	23/08/62	18/08/64	726	8
D5909	07/03/62	16/07/64	862	4
Average			789	

Notes:
1. D5902 was out of traffic for the longest period: 2.6 years.
2. Given the conversion work carried out on D5901 to allow fitment of the 'U' engine and then subsequent reversion to standard, this locomotive might be expected to have been out of traffic for the longest period; however, it was in fifth position at 2.3 years.
3. D5905 was out of traffic for the shortest period: 1.5 years.
4. Average time out of traffic: 2.2years.

With locomotives out of traffic for an average of 26 months, it might be expected that liquidated damage costs would have been imposed by BR based on a formula similar to that applied for late deliveries when first introduced, particularly for the period prior to commencement of rehabilitation. The English Electric orders state that no penalty was applicable for the period of commencement of rehabilitation work up to re-deivery to BR; this was either skilful negotiation on the part of EE, or surprising generosity on BR's part.

Chapter 11
RE-DELIVERY AND ACCEPTANCE

As the locomotives were returned to BR they were accepted at Doncaster Works, working trials with 250 to 290ton trains of coaching stock to New England, Peterborough. The locomotives were delivered to Doncaster starting with D5904 on 25 June 1964 and finishing with D5901 on 19 April 1965.

D5903, Manchester Victoria, 20 August 1964. En route from Newton-le-Willows to Doncaster Works for acceptance. After Manchester D5903 was routed via Woodhead, Penistone and Sheffield. *(Author's Collection)*

Below left: **D5909, Doncaster Works (Paint Shop Yard), 5 July 1964.** *(Barry Collins)*

Below right: **D5904, King's Cross, 1964.** *(Alec Swain [Transport Treasury])*

Full details are provided below:

Acceptance Tests.

Loco No.	Arrival at Doncaster	Known Test Train Dates	To Finsbury Park	Order returned to traffic
D5900	By 25/09/64	25/09/64	02/10/64	5
D5901	09/04/65	23/04/65		
		26/04/65	29/04/65	10
D5902	By 01/11/64	04/11/64		
		20/11/64		
		24/11/64	27/11/64	7
D5903	20/08/64	24/08/64		
		26/08/64	04/09/64	4
D5904	22/06/64	25/06/64	01/07/64	1
D5905	29/11/64	09/12/64		
		11/12/64	20/12/64	8
D5906	16/10/64	21/10/64		
		23/10/64	29/10/64	6
D5907	12/03/65	17/03/65		
		19/03/65	31/03/65	9
D5908	By 12/07/64	20/07/64		
		12/08/64	18/08/64	3
D5909	03/07/64	07/07/64	16/07/64	2

Notes:
1. The September 1964 edition of the *Railway Observer* reported that 'The "Baby Deltics"… are being run in on the test trip (3Z59) used for the 1750hp Type 3's from English Electric.'
2. D5908 failed on its test run on the 20 July 1964 with traction motor issues and oil leaks from the radiator and associated pipes and hoses. It returned to Doncaster light-engine, with the stock following behind hauled by WD 2-8-0 steam locomotive 90580.

Chapter 12
PERFORMANCE AND SERVICE PROBLEMS (1964–71)

12.1 Maintenance Contract

The following archive material describes the setting up of a contract for the maintenance of the Napier 'Deltic' engines following rehabilitation:

Memorandum to Supply Committee, 4 February 1965
Maintenance of 'Deltic' Engines in Type 2 Locomotives
1. Notwithstanding the introduction of a number of modifications the 'Deltic' 9 engine incorporated in ten English Electric Type 2 locomotives, delivered between April 1959 and July 1959, still proved unsatisfactory, and, as a result, English Electric have offered to install rebuilt engines free of cost to the Board. This offer, together with a proposal to enter into a contract with English Electric for the maintenance of the rebuilt engines, was approved by the Technical Committee.
2. The terms and conditions of the maintenance contract have now been settled with English Electric, whereby the Board will pay 25/- per engine running hour, subject to a review after 18 months. The agreement would be for a minimum period of 3 years and thereafter subject to 6 months' notice by either party.
3. The Chief Mechanical Engineer concurs in the terms of the arrangement, and the Committee's authority is sought to finalise the agreement on the foregoing basis.

Signed: Supplies Manager, Supplies and Contracts Department.

Supply Committee Meeting, 11 February 1965
Minute 307. Maintenance of 'Deltic' Engines in Type 2 Locomotives
The Committee discussed a memorandum dated 4th February 1965, from the Supplies Manager, in regard to a proposed contract for maintenance of 10 rebuilt engines which had been installed in English Electric Type 2 locomotives. After discussion, it was agreed that Mr. Ratter, in consultation with Mr. Houchen and Mr. Harrison should circulate a paper on the subject, ex-Committee.

Supply Committee Meeting, 11 March 1965
Minute 312. Re. Minute 307. Maintenance of 'Deltic' Engines in Type 2 Locomotives
After discussion the Committee agreed that a contract should be placed with English Electric for the maintenance of 10 rebuilt 'Deltic' engines. The contract should be for a period of 3 years and should provide for the training of selected Workshop staff in English Electric Works during the period of the contract to enable us to take over the maintenance ourselves.

English Electric/Vulcan Works Order No. 6732 (dated 16 July 1965)
This order is issued to cover the overhaul by D. Napier & Sons of 'Deltic' T9-29 diesel engines installed in the Type 2 at 4000 hour intervals in accordance with the maintenance agreement reached between BRB and [EE] Bradford on 12th April 1965. Payment to made monthly by BRB at the rate of 25/-d per engine running hour. [N.B. No contract duration period was specified].

Two other sources provide additional insights.

Webb (1982) described that before the final refurbished locomotive returned to service:

> … a maintenance contract covering the Napier 'Deltic' engines was drawn up, based on a rate per engine hour, the rate decreasing as the contract engine hours increased on two-yearly cycles. It was proposed that repair periodicity would initially be 4000 hours, and then 6000 hours, and finally 8000 hours up to mid-1968. The first engine repairs would start in mid-1966 and not more than two locomotives were to be out of service awaiting engines at any one time, regardless of whether this was due to regular overhaul requirements or through premature failure.

No mention is made in BRB Committee Meeting minutes about the increasing period between repairs over time. It may be, therefore, that the financial benefits of the increased periodicity was reaped by English Electric, albeit with BR benefitting from improved locomotive availability as a result of reduced time out of traffic for repairs.

A short article in *Modern Railways* (September 1966, pp505/6) entitled '"Deltic" Type 2s Redeemed' amplified the benefits of the scheme

> Contract maintenance, for the engines only, costs the railways a fixed sum per engine hour, which presumably breaks even or allows a small profit over an engine life of 4,000hr. Both sides benefit from this arrangement. Napier have a potential profit proportional to the life they can get out of the engine; and BR have a maintenance charge directly related to the amount of work the locomotives do.

12.2 Post Rehabilitation Issues
12.2.1 Failure of Lubricating Oil Pipes and Hoses
Very early on after rehabilitation, the 'Baby Deltics' were exhibiting problems due to loss of lubricating oil caused by broken pipe/hose connections and in turn provoking engine failures. A letter from BRB Chief Mechanical Engineer to English Electric dated 2 September 1964 illustrated a very early incident:

> *Casualty of Locomotive No. D5909 on 10.8.64 - Lubricating Oil Lost at Flexible Hose Connection*
> The CM&EE, Eastern Region, informed me that locomotive No. D5909 became a casualty at Stevenage on Monday 10th August whilst working the 5.22a.m. passenger train from Hitchin to King's Cross, causing a 37 minute delay.
> The flexible hose in the pipeline from the scavenge pump to the radiators came off a section of the steel pipe to which it is clipped causing a loss of lubricating oil.
>   After leaving the scavenge pump there is a tee connection to the thermostatic valve and from this connection oil is passed to the radiators through an upward curving steel pipe to a tee connection in the roof, where the oil flow divides [into] the two groups of radiator elements. The flexible hose is interposed between the thermostatic tee connection and the dividing connection in the roof. The intermediate section of steel piping, approximately three feet long is entirely unsupported and it is apparent that the swaying of the locomotive during running caused the intermediate section of steel piping to pull on the hoses, ultimately causing the lower end to pull out of the lower flexible hose near the thermostatic valve tee connection.
>   I have enclosed a sketch showing a stay and clip fitted to the lubricating oil pipe from scavenge pump to radiator which has been carried out on locomotive D5903; will you please fit this clip to the remaining locomotives to be delivered.

It was also suggested that inadequate clipping between the steel pipes and flexible hoses was a contributing factor.

The degree to which the lubricating oil circuit and pipe/hosing securing arrangements were modified during the rehabilitation process is not known. However, an internal EE memo dated 11 September 1964 stated that 'when these locomotives were "shopped" to Vulcan Works for rehabilitation, the engines of nine locomotives had been removed at BR depot [Stratford] and all the pipework "dumped" in a pile in the engine room.'

It was, therefore, virtually impossible to mimic the pre-rehabilitation piping arrangements, even if that was considered adequate. It was noticed, however, that the clipping arrangements, based on the 'dumped' pipework received, varied between locomotives (i.e. different clips, fitted in different positions, or, not at all). Worse still, the clipping arrangements proposed by BR (as per letter dated 2 September above) were found to be different from drawings produced and held by EE.

EE proposed a composite method of securing the pipework taking the best elements of the EE drawing and the BR proposal, and involving the use of duo-clips at the joints, used prior to rehabilitation, and stay bars. This approach appears to have been adopted with two sets of duo-clips and four stay bars sent to Finsbury Park for fitting to each locomotive already in service.

A letter from the BRB Chief Mechanical Engineer to EE dated 15 December 1964 highlighted further failures:

Two cases of failure occurred on Monday, 30th November, the first with locomotive No. D5909, and the second with D5908...

The trouble with locomotive D5909 was due to the flexible hose, which couples to the inlet elbow of the lubricating oil radiator on the 'A' side being blown off the elbow, resulting in a loss of oil. Locomotive D5908 was exactly the same thing, except that this was on 'B' side.

These two failures, together with similar failures on D5909 on 4 January and D5905 on 22 February 1965 brought into focus another potential cause for the lubricating oil circuit failures i.e. lubricating oil pressure fluctuations. Once again it was noted that these incidents followed extremely cold nights when the locomotives had been standing overnight at Hitchin depot, and that the failures took place approximately thirty minutes after starting up.

EE undertook some tests by inserting a pressure gauge at various points in the lubricating oil circuit; when the pressure gauge was positioned immediately adjacent to the scavenge oil pump, and on opening up the engine to full speed, the pressure fluctuated violently between 20 and 35psi several times per second.

Jubilee and Centoflex clips at the elbow joint were tested between 50psi and 300psi under fluctuating conditions both without failure i.e. significantly above levels experienced in traffic.

The elbow joint problem was addressed by two actions:

- The use of additional support for the hose to prevent potential slippage at the joint; in early 1965, supports, carrier assemblies and clips were manufactured for fitting to the two locomotives still at Vulcan Foundry (D5901/7) and for retro-fitment to locomotives in traffic (during major examinations). D5900/3/4 had been fitted by mid-June 1965.
- The use of grooves at the hose ends to assist with effective gripping of the pipework.

Instructions were issued to all concerned that the radiator shutters should be kept closed until the system was sufficiently warmed.

12.2.2 Radiator Leakages

As previously mentioned, the 'Baby Deltics' returned to traffic after rehabilitation with radiators in assorted condition.

Not surprisingly, the original 'A' radiator installed in D5909 failed during August 1964 after two months in service with leakage from the matrix and it was despatched to Marston for inspection.

The re-tubed radiator panels in D5904 failed in October 1964 after less than three months in service. It was an event which triggered severe consternation amongst English Electric managers to the extent that the following Period Report comments were sent to the Chairman on 28 October:

During the weekly period up to 23rd October the operating performance was satisfactory with the exception of one locomotive D5904. Although no failure occurred in service there were signs of further deterioration of the oil radiators as mentioned in previous reports and with no remedial action which could be effected at site the railway operators withdrew the unit from service on the 22nd October pending receipt of spare radiators. At the time of writing all efforts are being made to obtain a satisfactory pair of radiators from the suppliers.

As you are aware from site reports the question of leakage of oil radiators is causing us some concern and so far we have no hard and fast solution available at site except to return defective radiators to the supplier for rectification with no guarantee of continued satisfactory operation afterwards. D5904 has been out of service for the last five days awaiting a replacement set and whilst we have now been promised dispatch of a set from Vulcan today Wednesday we still have the problem of suspect radiators to deal with on the other four locomotives in service.

It is more than likely on receipt of our weekly report that the Chairman may ask the question of why a locomotive should be out of service for such a period awaiting replacement and on this regard it would be helpful to have Messrs. Marston's comments on the results of their examination so far.

The reaction of the EE Chairman is not known but an inter-Departmental response dated 29 October was circulated as follows:

We think it is fair to point out that the first two locomotives to be delivered, namely D5904 and D5909, were fitted with sets of original radiators [contrary to an earlier EE memo with respect to D5909] which had been cleaned and tested at Vulcan Works and to the best of our knowledge were suitable for a further period in service. When these two locomotives were on the point of being delivered it was discovered by metallurgical examination of a sample taken from a similar radiator that corrosion of the tubes had taken place and the radiators in the locomotives were likely to be similarly affected. The Sponsoring Engineer was consulted and it was agreed that in order not to delay delivery the two locomotives would be accepted for service, and the radiators would be changed at a BR depot at sometime in the future when they began to give trouble.

As it happened the radiators began to leak far sooner than was expected and those in locomotive D5909 were changed early in September. A replacement set for D5904 was despatched to Finsbury Park yesterday.

The reason for the delay in sending a replacement set for locomotive D5904 is that (a) the epoxy resin treatment of the water sections was not satisfactory and all available radiators at Vulcan Foundry were held up pending the receipt of re-tubed water sections from Marstons, (b) as mentioned above we had to find two replacement sets of radiators for locomotives in traffic far sooner than was expected.

As we understand it the position at the time of writing is that two locomotives D5909 and D5903 are suffering from slight oil leaks at the joints between the header tanks and the tube plates. The radiators fitted to D5909 are a spare set which although manufactured some years ago have, as far as we know, never seen any service. They have recently been subjected to the epoxy coating process. The set fitted to D5903 are radiators which suffered mechanical damage to the tubes and were completely re-tubed by Marstons only a few weeks ago. They would be tested in the presence of the BR Inspector and here again there should be no reason for joint leakage on what are, in effect, new radiators.

Given the history of radiator issues, the subject of using radiators from an alternative supplier was proposed by BR as early as January 1965. An internal EE Memo from the Chief Mechanical Engineer at EE Newton-le-Willows dated 29 January 1965 made the following comment on this proposal: 'we do not consider that there is sufficient evidence to justify discussing a new design of radiator' and this conclusion was passed onto BR on 1 February 1965. The BRB CM&EE at Doncaster responded in decidedly strong terms to EE on 16 February, and in slightly disjointed English, as follows:

I am disappointed that your Chief Mechanical Engineer has given careful consideration to my suggestion that we should jointly think about a new design of radiator for these locomotives, but does not feel that there is sufficient evidence to justify discussing a change, as to me the reverse would be the case. These radiators have been a constant source of trouble since their inception, and I can assure you that there is plenty of evidence to justify a change, and I am by no means satisfied that we are not going to get further trouble with the unmodified radiators in service.

My final comment on this problem is that the first time any trouble occurs on these Marston radiators fitted to the locomotives, I shall feel justified in taking independent action in closely examining the design with a view to installing another make of radiator without the help of your Chief Mechanical Engineer, if the locomotives fail in service due to radiator defects.

The internal EE response to the BRB letter was robust, but quite what was said directly to BR is not known:

It is true that these radiators have provided a lot of trouble since their inception; however, the evidence needed to justify completely new radiators of another supplier is that of continued failure for which a suitable modification cannot be found. It is extremely unfortunate that neither the Railways, ourselves nor Marstons have kept records dealing with the precise location of such failures. It has been established now that any radiator failure in future is to be put in Marston's test tank so as to obtain the exact location

and cause. This has only been applied to about three cases so far and of these two were joint failures which can be corrected… We have no reason to expect (*a*) recurrence of failures in joints corrected in this way.

Although we have not got exact evidence of earlier failures we are quite certain that the majority of these arose from the tube material which has now been corrected. Therefore, it is our viewpoint that there is insufficient evidence to justify wholesale change together with the extensive modification which would be necessary to the locomotive to accommodate alternative schemes.

Fifteen months later, on 29 April 1966, EE visited Norman Isherwood, Turton, Lancashire, for discussion regarding the potential supply of combined coolant and lubricating oil corrosion-protected radiator panels for use in the 'Baby Deltics'. These radiators had copper tubes with cooling fins and were designed with all the previous radiator problems in mind, being almost completely interchangeable with them in terms of fittings and connections. At the suggestion of English Electric, BR decided to proceed with purchase of one set at a cost of £927 10s per pair (net). An order was placed with Isherwood in January 1967. As far as is known, the Isherwood panels were never installed and were kept as a spare set!

In early 1966 a new problem manifested itself, due either to the failure of soldered joints around the radiator tubes in the header plate, or fracture of the tubes adjacent to the header plate. It was agreed in March 1966 by BR and Marston that

the cause was differential expansion between the brass tubes and the radiator steel frame, and that in spite of slip joints in the frame to cater for this contingency they were not functioning properly. It was decided to fit brass sides instead of steel to the matrix and improve the design of the slip joints. Between May and August 1967 ten locomotives sets of radiators were put through the makers' works and fitted to the locomotives. Maybe this work alleviated the radiator problems to such an extent as to preclude the need to deploy the Isherwood panels.

Radiator leakages were, however, an ongoing problem and panels were continually passing to and from the makers' works, or to EE for inspection, testing or repair work throughout the 1965-70 period and it was not unusual for locomotives to be in service with leaking radiators as evidenced by the frequently seen and obvious leakage marks below the bodyside radiator grilles!

12.3 Technical Report to the BRB (October 1967)

A Technical Report to the BRB dated 16 October 1967 included an Addendum entitled 'Diesel Locomotive Performance and Efficiency'. Commentaries on the various classes of locomotives operated by BR were provided; these provided a useful and succinct summary of the issues being addressed by the ER engineers. Details are given below:

Type 2: English Electric:
10 locomotives
These 10 locomotives are fitted with the 9-cylinder 'Deltic' engine. Repeated trouble with

this engine in the early years led to a complete refurbishing of the engines in 1964/5 by the manufacturer at their expense. The engines are maintained by the manufacturers under an agreement by which the Board pays for this maintenance on the basis of the engine hours run. Arrangements are being made for the BR Workshops to undertake this work when the present agreement expires in 1968. Availability has currently fallen to just below 80%, but is expected to recover to about 82% during early 1968 and be held at that level thereafter, being some 5% short of the National Plan level of 87%. The current casualty rate of around 12,000 miles per casualty is only expected to show a marginal improvement in the future.

Date Sheet 7C. Diesel Locomotive Performance data (position at 30/06/67):

1. *Availability (%)*. 85% end-1965, deteriorating to 82% end-1966 and 80% by mid-1967. Forecast to increase to 82% by end-1970.
2. *Reliability (Miles per Casualty)*. 10,000 miles per casualty end-1965, improving to 17,000 in mid-1966, deteriorating to 12,000 by mid-1967. Forecast to 'flat-line' at 12,000 through to end-1970.
3. *Principal Technical Problems.*

Item	Action Required or Taken
General Mechanical	
Auxiliary gearbox traction motor blower drives fracturing.	Spares being fitted and manufacturer examining the design from design and quality aspects.
Radiators splitting causing oil and coolant leaks.	Modified radiators fitted. One locomotive to be fitted with an improved type of radiator and dependent on the results with this is whether the whole fleet will be so modified.
Lubricating oil hoses becoming detached when under pressure.	Pipe ends modified and existing Centoflex clips replaced by Jubilee clips.

4. *Modifications*

Modification	Estimated spread of outstanding expenditure (£,000)			
	1967	1968	1969	1970
Authorised				
1. General Mechanical	0.09	0.25	0.25	
2. General Electrical	-	-	-	-
3. Engines	-	-	-	-
4. Main Generator	-	-	-	-
5. Traction Motors	-	-	-	-
6. Bogies	-	-	-	-
7. Heating Boiler	0.05	0.02	0.01	-
Total	0.14	0.27	0.26	-

Modification	Estimated spread of outstanding expenditure (£,000)			
	1967	1968	1969	1970
To Be Authorised				
1. General Mechanical	-	2.10	2.10	2.10
2. General Electrical	-	-	-	-
3. Engines	-	-	-	-
4. Main Generators	-	-	-	-
5. Traction Motors	-	-	-	-
6. Bogies	-	-	-	-
7. Heating Boiler	-	-	-	-
Total	-	2.10	2.10	2.10
Grand Total	0.14	2.37	2.36	2.10

12.4 'Deltic' Type 2s Redeemed?

The Class 23 locomotives were certainly much improved by the rehabilitation work, and apart from the oil leakage and radiator problems were now reliable locomotives. Availability for the refurbished locomotives was regularly 80-90 per cent and 100 per cent in some weeks.

Modern Railways (September 1966) included a short reference to the 'Baby Deltics' entitled '"Deltic" Type 2s Redeemed"; a slightly abridged version of this is reproduced below:

> The 10 English Electric Type 2 locomotives powered by the nine-cylinder 1,100h.p. version of the Napier 'Deltic' engine were the one blot on EE's diesel record with BR. After a long period of engine bothers, BR lost interest in the troublesome diesels and as the engines broke down sent them to Stratford to be put in store…..While talks went on between EE, BR and Napier on rehabilitation, the locomotives suffered badly from neglect and the attentions of the local vandals. In the end EE managed to avoid the stigma of one of their designs having to be re-engined and the so-called 'Baby Deltics' were taken to Vulcan Foundry to be fitted with modified engines. Because of the rigours of exposure at Stratford the locomotives had to be completely overhauled and returned to service virtually as new. The refurbished locomotives are now in their second year of service and the remedies applied to the small 'Deltic' engine seem to have worked. The only trouble experienced was the blowing-off of lubricating-oil hoses due to the high pressures produced when running up on cold engine oil; this was effectively cured by the use of a more positive hose connection. The engines are now running to the scheduled 4,000hrs between overhauls, and are being run for anything up to 100hr weekly, sometimes more, on the suburban trips out of King's Cross. Individual weekly mileages on this service have gone as high as 1,700.
>
> There remains one problem to solve: the perennial one of exhaust smoke. In this respect the small 'Deltic' seems a worse offender than most two-strokes. Opening the throttle in Gasworks Tunnel on the climb out of King's Cross produces a dense cloud of noxious oil-smoke which old-hand commuters say compares with the days of steam for unpleasantness, although in those days it was expected and windows remained closed.

12.5 BR Responsibility for Overhauling the 'Baby-Deltic' Engines

On expiry of the 'Baby-Deltic' post-rehabilitation maintenance contract in 1968 it was BR's intention to take over this work and also the work on the 'Deltic'

D18-25 engines in D9000-D9021. A training scheme was set up for BR Doncaster Works personnel at Napier's Liverpool Works and run so that, in the first instance, BR could phase-in the overhaul of the 'Deltic' T9-29 engines at Doncaster with effect from April 1968.

A considerable amount of expense, time and effort went into this scheme; however, the National Traction Plan published in November 1967 determined that the 'Baby-Deltic' fleet would be withdrawn from service commencing 1969 as part of a policy to eliminate numerically small and 'non-standard' locomotives. This edict effectively quashed the planned programme for 'Deltic' T9-29 engine repairs at Doncaster, in spite of the actions already underway.

The original intention was that the work and experience gained on the smaller T9-29 engine would pave the way for the second stage scheduled for 1969 when the larger 'Deltic' D18-25 engines used in D9000-D9021 locomotives would start to be Doncaster-maintained. In the event BR took over full responsibility of the D18-25 engines with effect from September 1969 as intended.

It is not known whether the original T9-29 maintenance contract was extended to reflect the phasing-out period of the 'Baby-Deltics' which was 'achieved' in March 1971 or whether BR had obtained sufficient expertise to support maintenance of the T9-29 engines in-house. Alternatively, the T9-29s may have been allowed to run without major attention (other than injector repairs at 2000hr intervals) until such time as the engines failed, at which point withdrawal was instant.

D5903, D5905, D5900 and D5904, Hitchin, 25 May 1967. (Frank Hornby [Colour-Rail])

Chapter 13
MILEAGES

13.1 Mileages Pre-Rehabilitation

Recorded mileages for the 'Baby-Deltics' up to 1 October 1962, together with subsequent calculated pro-rata mileages up to withdrawal for rehabilitation:

Mileages Pre-Rehabilitation.

Loco No.	Date Introduced	Mileage recorded at 01/10/62	Date taken out of service pre-rehabilitation	Pro-rata additional mileage	Calculated mileage pre-rehabilitation
D5900	22/05/59	148,460	25/01/63	12,610	161,070
D5901	22/05/59	146,620	28/01/63	12,934	159,550
D5902	01/05/59	109,040	16/04/62	0	109,040
D5903	17/04/59	113,650	14/02/62	0	113,650
D5904	24/04/59	128,360	01/05/62	0	128,360
D5905	08/05/59	133,210	13/06/63	27,592	160,800
D5906	08/05/59	145,810	05/10/62	539	146,350
D5907	15/05/59	159,500	13/11/62	4,742	164,240
D5908	29/05/59	141,580	23/08/62	0	141,580
D5909	19/06/59	106,350	07/03/62	0	106,350
Averages		133,260			139,100

Notes:
1. Average mileage per loco per day (introduction to 01/10/62) = 107.78
2. Pro-rata additional mileage: theoretical calculation based on additional time spent in traffic after 01/10/62, based on 107.78 miles per day achieved during the period between date introduced and 1 October 1962 and applied pro-rata thereafter.

13.2 Mileages Post-Rehabilitation

Recorded mileages for the 'Baby-Deltics' following rehabilitation as at 6 January 1967, together with subsequent calculated pro-rata mileages:

Mileages Post-Rehabilitation

Loco No.	Released into traffic post-rehabilitation	Order released post-rehabilitation	Mileage (after rehabilitation to 06/01/67)	Mileage Ranking	Date Withdrawn	Pro-rata additional mileage	Calculated mileage post-rehabilitation
D5900	02/10/64	5	107,970	5	30/12/68	94,520	202,490
D5901	29/04/65	10	88,160	10	06/09/69	114,440	202,600
D5902	27/11/64	7	102,950	7	23/11/69	119,690	222,640
D5903	04/09/64	4	116,340	3	30/12/68	94,520	210,860
D5904	01/07/64	1	124,070	2	20/01/69	96,640	220,710
D5905	20/12/64	8	104,510	6	14/02/71	149,890	254,400
D5906	29/10/64	6	101,930	8	30/09/68	85,320	187,250
D5907	31/03/65	9	89,330	9	20/10/68	87,340	176,670
D5908	18/08/64	3	111,600	4	09/03/69	101,500	213,100
D5909	16/07/64	2	125,510	1	07/03/71	151,300	276,810
Averages			107,240			109,510	216,750

Notes:
1. Average mileage per loco per day (re-introduction to 06/01/67) = 134.79.
2. Pro-rata additional mileage: theoretical calculation based on additional time spent in traffic after 06/01/67, based on 134.79 miles per day achieved during the period between release from rehabilitation and locomotive withdrawal applied pro-rata, subject to the following moderating factors:
3. Mileage rate after 30/09/68 reduced by 25 per cent reflecting the reduction in passenger duties performed.
4. Mileage rate for D5901/2/5/9 reduced by 50 per cent after 31/03/69 reflecting removal from front-line passenger duties.
5. D5901 withdrawal taken as the date transferred to Derby Research Centre, rather than its official withdrawal date.

13.3 Total Mileages

Full-life BR mileages (estimated):

Gross Mileages

Loco No.	Total Recorded + Calculated Mileage Pre-Rehabilitation	Total Recorded + Calculated Mileage Post-Rehabilitation	Total Calculated Mileage	Comments
D5900	161,070	202,490	363,560	
D5901	159,550	202,600	362,150	BR Mileage only
D5902	109,040	222,640	331,680	
D5903	113,650	210,860	324,510	
D5904	128,360	220,710	349,070	
D5905	160,800	254,400	415,200	
D5906	146,350	187,250	333,600	
D5907	164,240	176,670	340,910	
D5908	141,580	213,100	354,680	
D5909	106,350	276,810	383,160	
Averages	139,100	216,750	355,850	

Notes:
1. As a check on these calculated figures, it is worth mentioning some statistics quoted by R.M. Tufnell (1979):
   - D5901 achieved 252,000 miles in traffic with BR between mid-1959 and withdrawal in 1969, with fourteen reported casualties. Subsequent work with the Derby Research Unit increased this total to 491,000 miles.
   - The average mileage for the ten locomotives was 300,000 with an average miles per casualty figure of 12,900.
2. The recorded figures with my calculated additions to cover the unrecorded periods (with some admittedly very "rough-guess" assumptions of daily mileages in the post-30 September 1968 period) suggest an average full-life mileage of nearly 356,000. For the average figure of 300,000 to apply there would have had to have been a substantial, even extreme, reduction in the level of usage post 06/01/67.

Chapter 14
ACCIDENT AND FIRE DAMAGE

14.1 Accident Damage
No major incidents involving the 'Baby Deltics' have been found.

14.2 Fire Damage
14.2.1 BR Fire Reports
BR had serious problems with diesel locomotives catching fire during their early years. In an attempt to address the issue and the associated safety and cost implications, BR set up regional reporting of fires so that the reasons and severity could be ascertained on a class-by-class basis. This commenced in 1961 and over the following years the reports on 'Fires on Diesel Train Locomotives' produced by the Locomotive Performance and Efficiency Development Unit, Derby, contributed greatly to the understanding of the problem, quickly identifying potential solutions (e.g. locomotive modifications [e.g. re-routed pipework, spark guards], more effective fire-fighting equipment, improved maintenance and cleaning routines, etc).

The full listing of 'Baby Deltics' reported are given below:

Baby Deltic' Fire Incidents.

	Date	Source of Ignition	Material Involved	Degree of Damage	No. of Incidents
1961-64					
	None reported				0
1965					
D5906	28/12/65	B	4	S	1
1966					
D5903	28/07/66	B	1	NS	2
D5905	10/10/66	B	1	NS	
1967					
D5904	26/04/67	B	4	NK	2
D5904	07/11/67	B	2	NS	
1968					
	3 months-ended 31/03/68 (D5904 12/02/68, D5902 11/03/68 - see Notes)				?
	3 months-ended 30/06/68 (D5904 01/04/68 - see Notes)				1
	3 months-ended 30/09/68 (D5909 13/08/68 - See Notes)				1
	3 months-ended 31/12/68 (D5909 15/11/68, D5905 03/12/68 - see Notes)				3

Accident and Fire Damage • 159

	Date	Source of Ignition	Material Involved	Degree of Damage	No. of Incidents
1969					
	3 months-ended 31/03/69				0
	3 months-ended 30/06/69				0
	3 months-ended 30/09/69				0
	3 months-ended 31/12/69				1
1970					
D5909	10/11/70	B	2	NS	1
1971					
	None reported				0

Notes:

1. Abbreviations:
   Source of Ignition:
   A Brake block sparks.
   B Hot engine parts, including exhaust blows.
   C Electrical overloads or arcs.
   D Mechanical seizures or failures
   E Train heating boilers or burners

   Combustible material involved:
   1 Oil impregnated dirt or waste.
   2 Fuel or lubricating oil sprays or leaks.
   3 Electrical insulation.
   4 Not known.

   Degree of damage:
   S   Severe (necessitating Works attention).
   NS   Not severe.
   NK   Not known.

2. D5904 exhaust fire 12/02/68 between Baldock and Royston; attended to by local fire brigade.
   Memo from ER Movements Manager to BRB Chief Operating Officer, 3 April 1968
   Fire on Diesel Loco (D5904): Between Baldock and Royston 12 February 1968
   "Further to my letter of 19th February 1968. A fault occurred on the locomotive concerned which resulted in excess carry-over of lubricating oil into the exhaust system, which eventually became ignited."

   Three handwritten comments were attached to this memo:
   Maintenance? Or, These damned 2-strokes?

   Whilst I'm in sympathy with the well deserved adjectives, the makers allege that maintenance will prevent the build up of carbon and subsequent fires - this is by no means an isolated case nor is it insignificant - we've had lineside fires started by these monsters (on the same stretch of track) and must cure or withdraw them from service.

   Withdrawal will probably be complete by the end of the year.

3. D5902 exhaust fire 11/03/68 whilst working 06.41 Finsbury Park-Welwyn Garden City passenger; locomotive removed at WGC and attended to by local fire brigade.
4. D5904 exhaust fire 01/04/68 whilst working 17.15 Broad Street-Hertford passenger at Cuffley, extinguished by driver and locomotive continued journey.
5. D5909 exhaust fire 13/08/68 between Welwyn Garden City and Knebworth.
   Memo ER General Manager, York to the Ministry of Transport, 5 September 1968
   Locomotive D5909 on fire between Welwyn Garden City and Knebworth, 13 August 1968
   "Further to my report of 21st August 1968; this outbreak was due to engine sump oil leaking out from the exhaust manifold whilst the engine was idling and subsequently vapourising and igniting when the engine was subjected to loaded conditions. This is a known design shortcoming but this class of locomotive is due to be phased out of service in the near future."
6. D5909 fire 15/11/68 whilst working 0715 Royston-King's Cross at Woolmer Green; extinguished by driver.
7. D5905 fire 03/12/68 whilst working 0715 Royston-King's Cross at Hadley Wood.

Chapter 15

OPERATIONS: A HIGH-LEVEL SUMMARY

15.1 1959-1963

15.1.1 Overview

The 'Baby Deltics' were delivered to 34B Hornsey shed during April-July 1959 for use on passenger work, including inner- and outer-suburban Great Northern services from King's Cross and Broad Street, Cambridge express duties from King's Cross and empty coaching stock duties, together with some freight work during off-peak periods.

Occasional forays were made to Moorgate via the London Transport Metropolitan Lines and trip-cock apparatus was fitted to enable this; however, excessive fume issues in the tunnels resulted in the rostering of 'Baby Deltics' to such services being avoided or banned altogether except in emergency situations.

Their delayed introduction was predominantly due to the need to apply weight-saving initiatives to achieve an 18ton axle-load which would have allowed them to work on inter-regional transfer freight duties to the Southern Region. Ultimately this proved unsuccessful and the 'Baby Deltics' sphere of influence was restricted to north of the Thames with a heavier bias toward passenger traffic than originally intended. Of the eight Summer 1960 Monday-Friday diagrams only three included a freight trip during the off-peak with five including empty stock workings, between King's Cross and Bounds Green or Holloway.

In April 1960 the fleet was re-allocated to the new purpose-built diesel depot at 34G Finsbury Park although their duties remained unchanged with all turns during the first four years covered by 34D Hitchin men.

Despite (or because of) their operational restrictions the 'Baby Deltics' achieved the highest mileages of the Finsbury Park Types 1 and 2 diesels. David Percival (*Railway World*, May 1971) listed the average mileage for the various locomotive types during June 1961 as follows: 'Baby Deltics' 3,891, Brush Type 2s 2,539, BR/Sulzer Type 2s 743, English Electric Type 1s 1,366 and BTH/Paxman Type 1s 1,269. These mileages clearly reflected the activities undertaken with the 'Baby Deltics', at one end of the spectrum, concentrated on the longer-distance outer-Suburban services and Cambridge expresses, compared with the BR/Sulzer Type 2s heavily involved in slow short-distance cross-London transfer freight duties.

Monthly BTC 'Diesel Train Locomotives: Availability and Utilisation' reports for late 1962 recorded that 93 per cent of 'Baby Deltic' duties were on passenger work (Monday-Saturday).

15.1.2 The King's Cross Suburban Diesel Scheme

The transition from steam to diesel in the King's Cross area was the subject of an excellent article by Brian Perren in the May 1962 edition of *Modern Railways* entitled 'The King's Cross suburban diesel scheme'; with only relatively minor abridgement and some abbreviation modifications, the article is reproduced here:

The Great Northern (GN) Line's use of both diesel multiple-unit (DMU) trains and separate diesel locomotives for suburban traffic was the first adoption in this country of the formula followed by the Netherland Railways, where the obvious advantages of multiple-unit working for passenger services are complementary to the mixed traffic flexibility of separate locomotives. Employment of DMUs is restricted to a minimum number of units required for the basic passenger service; most of the additional peak services required are covered by locomotive-hauled stock worked by locomotives on comprehensive rosters which include a large amount of time on freight work outside of the passenger peaks. Thus a good day's work for the DMU trains on passenger services from approximately 6a.m. to midnight, and a 20hr day on freight and passenger work for the locomotives fully exploits all motive power equipment. The King's Cross area is particularly suited to this type of arrangement.

Before dieselisation, the basic suburban passenger traffic pattern in the King's Cross area comprised two types of train. There was an off-peak hourly stopping service from King's Cross or Finsbury Park to Hatfield or Welwyn Garden City and over the branch to Hertford North, with additional trains in the peak to or from Broad Street and over the Metropolitan Widened Lines to Moorgate. Outer suburban passengers were provided with an hourly service to Hitchin, with certain of these trains running onto the Cambridge branch, and four Cambridge buffet-car expresses calling at Welwyn Garden City, Stevenage, Hitchin, Letchworth and Royston… On the freight side, the GN Line operated daily some 20 to 25 return trips from Ferme Park to the Southern Region over the difficult cross-London route via the Widened Lines, Farringdon, Snow Hill and Blackfriars, together with transfer trains to Temple Mills, Ripple Lane and Acton via Canononbury. In addition, empty main-line stock had to be worked from sidings at Hornsey and Holloway to the terminus at King's Cross. Four types of steam power were employed on these jobs: B1 4-6-0s and L1 2-6-4Ts on the outer-suburban trains; L1 2-6-4Ts and N2 0-6-2Ts on the inner-suburban and empty stock trains; and J50 0-6-0Ts on the cross-London freights. However, the work of these four classes was limited by certain route availability restrictions over the Widened Lines which precluded the use of L1s beyond King's Cross to Moorgate and N2s over the route from Farringdon to the Southern Region. Moreover, engines working beyond King's Cross had to be fitted with LTE trip-cock equipment.

A new timetable introduced with diesel traction in the summer of 1959 substantially improved the passenger service from King's Cross both in frequency and speed. A basic half-hourly off-peak interval service to Hertford and Hatfield and hourly to Welwyn Garden City was provided by 28 Cravens two-car DMUs operating in two-, four- and six-car formations as required. In the peak period, fast services avoiding calls at certain stations were provided by locomotive-hauled trains. Apart from minor adjustments of stops in the peak periods, the basic service has not been changed since its inception.

Unfortunately, the full benefits could not be achieved, so far as the locomotives were concerned, when the new scheme first started. The diagrams had been prepared so that the locomotives would cover a return freight trip from Ferme Park to the Southern Region at night, work an up morning suburban train to King's Cross and the empty stock back to Finsbury Park sidings, then make a second freight trip to the Southern Region before heading an evening peak train out of the city and have an interval for daily maintenance inspection before the routine recommenced with a night freight working. (N.B. The South London freight service from Ferme Park cannot be operated in either direction during the passenger peak period.)

Originally, three different makes of Type 2 diesels were allocated to the GN Line - 10 English Electric 1,100hp 'Baby Deltics', 10 North British Loco. Co. (NBL) 1,100hp and 20 Birmingham R.C.W. (BRCW) 1160hp locomotives. On delivery it was found that none of these three designs would satisfy the requirements of the Southern Region's Civil Engineer, thereby precluding their use on South London freight workings.

A general tidying up of the Type 2 resources in the King's Cross area, saw the NBL and BRCW Type 2s re-allocated to Scotland with the introduction of 1365hp Brush Type 2 locomotives.

Despite the considerable time, effort and money being expended on reducing the overall weight of the 'Baby Deltics' it proved impossible to reduce the axle-loading to meet Southern Region restrictions (18tons maximum). Even the 'A' axles of the Brush Type A1A-A1A design exceeded this limit.

As a result the original scheme to employ any of the Hornsey Type 2 designs on freight transfer trips to the SR, as part of their off-peak duties, had to be substantially modified and the continued use of Hornsey-based J50 0-6-0T steam locomotives for a further two years was necessary until a suitable diesel type was found acceptable to the Southern Region civil engineers; in fact, some of the rather antiquated J50s (built between 1914 and 1939) received General Works repairs during this period to maintain continuity of operations. One J50, 68891 (built 1914 and withdrawn from the King's Cross area in July 1961) managed a life of 47 years, virtually five times as long as the 'Baby Deltics' (excluding D5905/9).

After much deliberation, the SR approved use of the BTH 800hp Type 1s (Class 15) on the cross-London freights from Ferme Park to the Southern Region (Hither Green, Norwood, New Cross Gate) via Farringdon and Snow Hill Tunnel; this in itself was not the solution due to the low power of the Type 1s and in any case there were insufficient of these locomotives available to allow the complete eradication of steam.

Ultimately a solution was found in the form of the BR/Sulzer Type 2 1160h.p. locomotives (deliberately selected from the lighter Type 11/1A batch, later Class 24/1, with lower capacity fuel and water tanks, and lighter train heating boiler); twenty-five (D5050-72/94/5) were transferred to Finsbury Park in the period January-July 1961 to cover these duties, thereby facilitating the demise of the J50s at the southern end of the Great Northern over the same period. Some Brush Type 2 diesels were transferred away in exchange to ensure effective utilisation of the BR/Sulzers on both suburban passenger and transfer freight work. Interestingly the weight diagrams for the BR/Sulzer Type 2 locomotives allocated to Finsbury Park still showed axle-loadings marginally in excess of 18tons!

Continuing with Perren's article:

The allocation of Type 2s to Finsbury Park depot and the duties they perform is now (*in 1962*) as follows: *See table below.*

Although the 75 engines ... are all in the Type 2 category, differences in design have made it necessary to allocate each of the three makes to specific traffic. Since the BR/Sulzer is the only type permitted to South London their use is confined mainly to freight transfer trips and suburban peak work; the ten 'Baby Deltics' were sufficient to form a small sub-allocation based on Hitchin depot and therefore cover outer-suburban trains to Hitchin and beyond. Because of their additional power the Brush locomotives are very suitable for smart working of the heavy empty stock of main-line trains, particularly sleeping car sets, over the sharp climb out of the terminus to Holloway and Hornsey sidings.

Engine	No. Allocated	Duties
Brush 1365hp	40	Empty stock workings to and from King's Cross; peak suburban; local freight; transfer freight to Acton, Temple Mills, Ripple Lane.
BR/Sulzer 1160hp	25	Cross-London freight to the SR; peak suburban; empty stock.
English Electric 1100hp	10	Cambridge and outer suburban trains; local freight; empty stock from King's Cross.

Perren provided examples of rosters, one for each of the three classes. The 'Baby Deltic' roster, and associated commentary, is given below:

One-Day Roster for English Electric 1,100hp Diesel.

Terminal	Arrive	Depart	Miles
Hitchin	-	5.58a.m.	-
Royston	6.31a.m.	7.05a.m.	13
King's Cross	8.22a.m.	9.30a.m.	44
Baldock	10.37a.m.	12.57p.m.	37
King's Cross	2.09p.m.	3.05p.m.	37
Ely	4.54p.m.	5.30p.m.	73
King's Cross	7.28p.m.	8.30p.m.	73
Baldock	9.44p.m.	-	37
Hitchin Depot	-	-	5

Each diagram includes a period for daily examination, which can be made either at the yard adjoining King's Cross passenger station, at Hornsey or at Hitchin depot. Seven-day, 14-day, monthly and periodical examinations are covered at Finsbury Park maintenance depot. At present the Brush diesels are running between 3,500 and 4,000 miles per month, the 'Baby Deltics' approximately 4,500 and the BR/Sulzers between 1,500 and 2,000. There are also 10 English Electric Type 1 1,000hp mixed-traffic diesels and six AEI 800hp units engaged on local freight work in the King's Cross area.

15.1.3 Trials and Tribulations

At various times during 1960/61, failure of the Napier engines in the 'Baby Deltics' caused extreme difficulties in rostering Great Northern services. On at least two occasions there were reports of them being relegated to freight duties or e.c.s. duties only, and there were also periods when low usage of the fleet was attributable to planned engine modifications. From time to time Brush Type 2s were loaned to Finsbury Park from other depots to offset severe motive power shortfalls. Following modifications, the fortunes of the fleet was revived and in 1961 (at least) some of the 'prestige' Cambridge buffet services were extended to Ely to utilise the previous lay-over time at Cambridge and to eliminate a Great Eastern DMU working between Cambridge and Ely.

However, matters came to head once more during late 1961/early 1962 when a spate of high-profile failures on Royston and Hitchin trains apparently caused the Great Northern management to relegate the class as a whole to local freight and yard shunting work (overall motive power availability permitting, of course), necessitating the use of Finsbury Park allocated (Hitchin sub-allocated) freight locomotives or imported motive power to cover, as reported by various railway periodicals:

> With only three 'Baby Deltics' available in mid-December (*1961*), Brush D5692, D5804/5/14/5 were on loan to Finsbury Park from Darnall. (*Railway Observer*, February 1962)

> Other unusual motive power to be encountered on Great Northern suburban duties in May (*1962*) were 1,000hp English Electric Type 1s, which also succeeded in covering most of their freight commitments... The cause of the trouble was the continuing unreliability of the 1,100hp English Electric Type 2 'Baby Deltics', of which no more than two have been observed in service on any one day recently. Their use on passenger work has been in jeopardy for a considerable time and they are now officially restricted to freight

and ECS workings, although one of the serviceable locomotives has been noted on the 4.05p.m. Class B Hitchin-Huntingdon and return and the 7.28p.m. local from Hitchin to Baldock. (*Modern Railways*, July 1962)

Several Sheffield Brush diesels have again been noted working local trains in the Kings X area, presumably due to another acute shortage of 34G units. This is underlined by the fact that some 800hp diesels based on Hitchin, for the local goods workings, have also been regularly used on suburban work for the past two weeks. D5680/1/2/3 from Darnall have also been noted. (*SLS*, June 1962)

D5901, King's Cross, 1959. (Grahame Wareham Collection)

D5900 and A1 Pacific 60145 *Saint Mungo*, King's Cross, Undated. (Jack Ray [courtesy Stuart Ray])

N.B. The referral to '800hp diesels based on Hitchin' should read 1,000hp diesels, although it should be noted that Finsbury Park BTH Type 1s were occasionally forced onto suburban duties during periods of serious motive power difficulties.

D5680-3 have arrived at Hitchin in exchange for D8021/3/4 which have gone to Darnall (*May 1962*). The transfer has virtually sealed the fate of the 'Baby Deltics' as far as passenger work is concerned. (*Railway Observer*, July 1962)

As discussed in Section 9.7, from February 1962 members of the Class were progressively taken out of use and stored, with all removed from traffic by May 1963. BR's official performance and availability statistics recorded that, with effect from January 1963, the remaining 'Baby Deltics' had no diagrammed duties and were only used on an 'as required' basis with the duties of the 'Baby Deltics' progressively taken over by Brush Type 2s.

Somewhat ironically, however, the following comment was reported in the March 1963 edition of *Modern Railways*:

In the London area, the fallibility of diesel train-heating equipment was underlined by an influx of B1 4-6-0s to empty stock workings into and out of King's Cross. The three serviceable 'Baby Deltics', D5900/1/5, had their restriction on regular passenger train haulage temporarily lifted, apparently because their train-heating boilers are more reliable than others in the diesel fleet.

Operations: A High-Level Summary • 165

Above left: **D5907 sandwiched between 60032 *Gannet* and 60055 *Woolwinder*, King's Cross, 2 September 1960.** (I.S. Carr [David Dunn Collection])

Above right: **D5903, King's Cross, Undated.** The discs indicates a local/suburban service. (Transport Treasury)

Below left: **D5905, New Southgate, 1961.** Ancient and modern; relatively new 'Baby Deltic' with two sets of 'Quad-Art' (4-coach articulated) coaching stock. (Grahame Wareham Collection)

Above right: **D5903, New Southgate, 23 October 1961.** Discs set for suburban passenger duties. (R.G. Warwick [David Dunn Collection])

Above left: D5907, north end of Barnet Tunnel approaching Oakleigh Park, 23 October 1961. Evening commuter service to Baldock or Royston. (R.G. Warwick [David Dunn Collection])

Above right: D5900, Welwyn Garden City, 23 December 1961. (R. Puntis [Grahame Wareham Collection])

Below left: D5907, south of Hitchin, 12 May 1962. Up engineer's train of track panels. (Charlie Verrall)

Below right: D5908 near Hitchin, 12 May 1962. 09.28 Ashburton Grove-New England Class 9 unfitted freight. (Charlie Verrall)

D5904, Hitchin, 7 June 1960. This photograph was included in *Classic Diesels and Electrics* Issue No. 3 (December 1997/January 1998); the caption read "D5904 propels a withdrawn steam locomotive at an unknown location shortly after introduction." Slightly wide of the mark because, in reality, the 'Baby Deltic' was awaiting a tow from Hitchin to Stratford DRS for an engine change (undertaken between 9 and 25 June 1960) and the two nearest identifiable steam locomotives (N2 tanks 69583 and 69543) survived until withdrawn in September 1962 and September 1961 respectively. (Nigel Petre)

D5908, Huntingdon, 7 September 1961. (Eric Sawford [Transport Treasury])

15.2 1964-1969

15.2.1 Overview

Following rehabilitation in 1964/65 and up to 1969, the 'Baby Deltics' returned to the Great Northern being heavily involved, as before, with King's Cross e.c.s. work, inner- and outer-suburban duties and express services to Cambridge and, occasionally, Peterborough. It appears that D5904 just squeezed in the odd 'Cambridge Buffet Express' duty before these were discontinued, in name, in early July 1964 (although the services continued to run 'anonymously'). Their strong association with the Hitchin sub-depot continued.

Locomotives built with headcode boxes were capable of displaying four-character codes to describe the trains they were hauling. Pre-refurbishment this was not possible as far as the disc-fitted 'Baby Deltics' were concerned; however, following rehabilitation in 1964/65, these locomotives were retro-fitted with indicator boxes and were, therefore, capable of displaying the four digits.

Headcode details relevant to the 'Baby Deltic' passenger operations are given below. These codes were applicable in the King's Cross area from 1 January 1961 and were used for many years thereafter.

First digit:
0   Light Engine
1   Express Passenger
2   Ordinary Passenger
3   Empty coaching Stock

Second digit (Area letter):
B   King's Cross

Third digit:
6   To/from King's Cross
7   To/from Moorgate
8   To/from Finsbury Park
9   To/from Broad Street

Fourth digit:
0   To/from New Barnet
1   To/from Potters Bar
2   To/from Hatfield
3   To/from Welwyn Garden City
4   To/from Hitchin
5   To/from Baldock or Royston
6   To/from Cambridge
7   To/from Gordon Hill
8   To/from Cuffley or Hertford.

Thus 2B65, heavily illustrated in this book, was a King's Cross to Royston or Baldock working and 1B66 applied to the Cambridge expresses. Main-line empty coaching stock into and out of King's Cross regularly used the code 3B59.

As already discussed, the timetable operated in the King's Cross area between mid-1959 and up to 1969 was based on maximising the use of DMU trains for the basic passenger service supported by locomotive-hauled trains during the peak-hour. Deployment of the locomotives on freight working during the off-peak period enabled a relatively high utilisation of all traction types, locomotives and DMUs, and this continued to be the case when the 'Baby Deltics' were re-introduced in 1964.

To increase flexibility of the usage of the 'Baby Deltics' on their return to traffic, a number of diagrams included passenger turns to and from Broad Street were covered by King's Cross and Hornsey men who had had no previous experience of the type, together with an increased number of King's Cross empty stock workings.

During the late summer of 1966, trials were carried out with two 'Baby Deltics' on some of the Cambridge expresses previously covered by a single locomotive with the aim of improving time-keeping; the 17.40 Cambridge-King's Cross on August 29-31 featured D5903/8, and a pair were noted several times during the following month.

During the latter part of this period, however, a number of factors resulted in the reduced use of locomotives ultimately resulting in the transfer away of the BR/Sulzer (Class 24) locomotives from freight duties in the King's Cross Division and the demise of the 'Baby Deltics', although this demise proved to be somewhat protracted. The three key factors were:

- The implementation of the National Freight Train Plan commencing 1966.
- The cascade of 'high-density' DMUs, predominantly from the Great Eastern, which usurped a significant proportion of the locomotive-hauled peak hour services from May 1969.
- An increasing number of main-line train sets turned round in the King's Cross platforms, without removal to carriage sidings and thereby reducing pilot duties.

15.2.2 Implementation of the National Freight Train Plan

The first stages of the BR National Freight Train Plan (NFTP) were brought into operation at the start of the new working timetable on 18 April 1966. This initiative had the dual objectives of reducing costs (including a degree of marshalling yard rationalisation and the concentration of specific traffic types (express block trains, wagonload traffic, etc) onto fewer dedicated routes), and improving service flexibility and reliability. These changes were brought about against the backdrop of increased volumes of block-train traffic, the introduction of coal concentration schemes and a general decrease in wagonload traffic.

Implementation of the NFTP radically altered the pattern of freight working in the King's Cross Division. With the objective of concentrating fast freight traffic on one route, the East Coast route became the principal route for express movements. Express freights from Temple Mills to the North, which previously took the GE route via Whitemoor, were transferred to the East Coast main line. Some Freightliner trains from Stratford terminus were similarly re-routed, together with a number of company trains from Thameshaven and the Southern Region; these reached the King's Cross Division via the North London line, Canonbury Junction and Finsbury Park.

Much of the King's Cross and Hitchin area wagonload traffic was routed via Temple Mills to combine with Great Eastern volumes for onward movement; this change inevitably was to the detriment of the Ferme Park and New England (Peterborough) marshalling yards. Ferme Park ultimately closed and New England was significantly down-sized.

Also under the NFTP the heavy flows of coal from the East Midlands previously using the Great Northern route from Mansfield through Lincoln, Spalding, New England, Ferme Park and the Widened Lines to the Southern Region was concentrated on the Midland route through Toton and Cricklewood.

As a result of these changes, the number of locomotives required for freight duties was substantially reduced and associated infrastructure rendered redundant, with wagonload traffic around King's Cross Goods and transfer traffic to the Southern Region particularly heavily affected. The impact in locomotive terms was the mass transfer of BR/Sulzer Type 2s (D5050-72/94/5) from Finsbury Park to the London Midland and Scottish Regions in the period July-October 1966. The 'Baby Deltics' and the Brush Type 2s (Classes 30 and 31) continued on the peak-hour locomotive-hauled suburban traffic, but the loss of so much local freight traffic inevitably reduced the utilisation of the 'Baby Deltics' leaving them highly vulnerable to subsequent motive power developments.

Residual freight traffic handled by the 'Baby Deltics' in the late 1960s included oil to the distribution depots in the King's Cross Division, naptha feedstock from Fawley to the Cadwell Gasworks (Hitchin) which regularly involved pairs of locomotives, coal distribution, and occasional deployment on the rubbish train from Ashburton Grove to Blackbridge tip (near Wheathampstead on the former Dunstable branch).

The proposed withdrawal of the BTH/Paxman Type 1s which began in early 1968 made it necessary to find alternative motive power for the three Finsbury Park Type 1s (D8231-3) on the Ashburton Grove rubbish. The heavily loaded early morning service in particular dictated the use of two locomotives in multiple to provide adequate braking power. In the spring of 1968 the turn was re-diagrammed for a single Brush Type 2 but a pair of 'Baby Deltics' was tried out on March 20; the 06.33 from Ashburton Grove and 08.42 return empties were powered by D5901/4, the former then taking the more lightly loaded 11.15 down service. Sanding equipment had been removed from the 'Baby Deltics' by this time and this would have made them less than ideal for coping with the 1 in 52 gradient between Welwyn Garden City and Ayot, certainly when operating solo. Space restrictions at Blackbridge may have also precluded the straightforward use of pairs of 'Baby Deltics' and Brush Type Type 2s.

In the event, only D8233 was withdrawn and D8231/2 were re-diagrammed to work the rubbish train until the end of 1970, at which point they were transferred to Stratford. With only two Type 1s available some substitutions were necessary to cover for maintenance repairs and so the pairing of a BTH Type 1 with a Brush Type 2 was not uncommon. The pairing of a 'Baby Deltic' and a BTH Type 1 was less common, but did occur, for example D5905 and D8232 on the early morning trip on both 3 and 21 November 1969.

15.2.3 'High Density' DMU cascades to cover Peak-Hour Operations

Twenty-four Cravens two-car DMUs, initially used on the Midland & Great Northern line, were sent to the King's Cross

Division in 1959 to work the 'new' dieselised service. In reality these 'low-density' low-powered (115 seats, 300hp) units were unsuitable for suburban services, particularly in peak hours; when operated in two- or three-set formations, the duplication of van space and toilet facilities resulted in a proportionate shortage of seats compared with locomotive-hauled suburban compartment stock. In addition, the small number of external doors on the DMUs extended dwell times at busy stations.

The Cravens fleet was sufficient to cover the basic daytime service but locomotive-hauled compartment stock was used to cope with peak-hour traffic. Although these trains provided more seats per train, they progressively become unattractive and obsolete on a modern railway.

In 1969 the opportunity was taken to cascade 'high density' DMUs displaced from other areas as a consequence of reduced traffic volumes or electrification schemes.

Four Derby-built three-car 600hp 'high-density' Class 116 DMUs (262 seats) were transferred to the King's Cross Division in June/July 1968, followed by 20 Derby-built three-car 952hp 'high-density' Class 125 DMUs (266 seats) in 1969/70 (16 sets in May 1969, 1 in June 1969, 3 in July 1970). The Class 125s came from the Great Eastern following electrification of the Lea Valley line; their 'high-density' design and higher power made them eminently more suitable for inner-suburban peak services. The progressive introduction of these units saw the demise of many of the locomotive-hauled suburban trains, with the 'Baby Deltics' the main focus for displacement. Six 'Baby Deltics' were withdrawn between September 1968 and March 1969 and a further two in November/December 1969. Any remaining off-peak freight services at this point could more than adequately be covered by Brush Type 2s.

The Cravens units continued on the off-peak services and Brush Type 2s operated the remaining peak-hour locomotive-hauled suburban services, strongly supported by the Class 116/125 DMUs, until November 1976 when the full Great Northern inner-suburban electric services to Welwyn Garden City and Hertford North commenced utilising Class 313 EMUs. Services to Moorgate via King's Cross were withdrawn permanently from the same date and the use of locomotive-hauled compartment stock ceased.

15.2.4 Reduction of Station Pilot and e.c.s. Duties

From the mid-1960s there was a move to reduce station-pilot duties by the quick-return of southbound services into King's Cross by the turn-round of stock northbound without recourse to empty coaching stock movements to and from carriage sidings for servicing or stabling. This process inevitably had an adverse impact on 'Baby Deltic' usage.

15.2.5 Snapshots of 'Baby Deltic' deployment (1964-69)

The following quotes provide a potted history of the life and times of the 'Baby Deltics' after refurbishment:

D5902/6 have now arrived back after modification and overhaul and have joined D5900/3/4/8/9 on outer-suburban workings… and at present appear to be quite reliable performers. (LCGB, January 1965)

… their performance and reliability seems to be about on a par with the Brush Type 2's. (LCGB, June 1965)

… their use has continued on the outer-suburban workings and all ten have been observed at work during the month (*September*). Even so, their availability seems to have dropped a little as D5622 was on the 07.06 Royston-King's Cross on 24/9 and D5623 on the 06.44 ex-Royston on 29/9. (LCGB, November 1965)

On the suburban side, the week commencing 29/11 [*1965*] saw a virtual disappearance of the 'Baby Deltics', although it is of course too early to say if this is a temporary or permanent absence. (LCGB, January 1966)

The disappearance of the 'Baby Deltics' turned out to be only temporary, possibly due to minor modifications, and they now seem to be all back in traffic. (LCGB, February 1966)

Despite the new arrivals (*four DMU sets*), there appears to be little reduction in the appearances of 'Baby Deltics' which in the past have proved themselves at least as reliable as the Brush Type 2's. (LCGB, October 1968)

No further three-car DMU's have arrived during the month and 'Baby Deltics' are still well in evidence on both inner- and outer-suburban workings and empty stock duties; it may well be that the lack of versatility of multiple units is causing the authorities to think again about the wholesale replacement of 'separate' power units. (LCGB, November 1968)

Punctuality [on the GN] has gone into severe decline during November and nowhere is worse than on the King's Cross suburban services, which usually have an exemplary record… Despite rumours to the contrary, the 'Baby Deltics' are continuing to flourish: D5900/2-5 all having been noted in traffic during the last week of November. The replacement 3-car DMU's are little in evidence. (LCGB, January 1969)

The Class 23 locomotives… due for withdrawal by the middle of 1969, are still very much in evidence on e.c.s. workings in and out of King's Cross, while a number are regularly employed on suburban trains to and from Broad Street. (*Railway Observer*, January 1969)

5901/2/5/8/9 have been noted working. (LCGB, March 1969)

5905 was noted working residentials [on 20 February 1969], so this Class is still operative, although its days must be numbered as 30A three-car DMUs are working training trips from Stratford to Baldock and the bulk of Stratford's twenty sets are expected to be transferred in as soon as the Lea Valley electrification [Stratford-Tottenham Hale-Cheshunt] is inaugurated. Their arrival may well end all loco-hauled suburban workings in the terminus, particularly as there are rumours that the Cambridge buffet services may be transferred to the Liverpool Street route. (LCGB, April 1969)

It is expected that the GE line DMUs will enter service on the GN on 4th May, following the switch-on of the Lea Valley Electrification scheme. This will leave only eight sets of compartment stock in use, mainly on outer-suburban workings. (LCGB, June 1969)

Electrification of the Lea Valley line from Coppermill Junction, between Clapton and Tottenham, to Cheshunt… was inaugurated for public passenger-carrying trains on May 5 [1969] … A residual diesel shuttle service has linked Tottenham Hale and Stratford from May 5. (*Railway Magazine*, June 1969)

As a result of electrification of the Lea Valley line from May 5, diesel multiple-units formerly used there have been transferred to the Great Northern suburban service, enabling the remaining 'Baby Deltics' to be withdrawn [sic]. (*Railway Magazine*, August 1969)

The Lea Valley electrification did not cause the complete demise of the 'Baby Deltic' locomotives as was expected… at least D5901/2 are still in service on suburban workings. (*Railway Magazine*, November 1969)

5909 worked 13.04 King's Cross to Cambridge on 14/11 [1969] and the three remaining members of this class are not infrequent visitors to Cambridge. (LCGB, February 1970)

D5905, King's Cross, 28 March 1969. 2B68 Sandy-King's Cross. (Peter Foster)

Above left: **D5901, King's Cross, 23 August 1969.** Within two weeks of transferring into Departmental service at the Research Centre, Derby and a very different lifestyle. (Keith Holt [The KDH Archive])

Above right: **D5902, King's Cross, 21 May 1969.** Headcode 2B65 was the generic headcode for all Baldock or Royston services. (Bill Atkinson)

Below left: **D5903, King's Cross, 12 March 1968.** Northbound e.c.s. The DLRC for D5903 indicates an Unclassified Works repair at Doncaster Works shortly after this photograph was taken with release on 20 March 1968. (Graham Clark [via Alan Monk])

Below right: **D5900, King's Cross, Undated.** General Manager's Saloon. (M.F. Best [David Dunn Collection])

Operations: A High-Level Summary • 173

Above left: **D5904 and D5678, King's Cross, 14 February 1967.** The first 'Baby Deltic' to receive full yellow ends only a few days prior to this photograph being taken. (David Percival)

Above right: **D5907, King's Cross, 1 May 1968.** (Nigel Petre)

Below left: **D5901, King's Cross, 12 July 1968.** 09.30 King's Cross-Hitchin service. (Jim Binnie)

Below right: **D5902, Moorgate, 29 August 1969.** A fairly rare shot of a 'Baby Deltic' on the Metropolitan 'Widened Lines'. (David Percival)

Above left: D5908. Hertford North, 12 June 1968. Service from Broad Street. (N.L. Cadge [Grahame Wareham Collection])

Above right: D5901, Hadley Wood, 14 August 1969. Between the North and South Tunnels. (George Woods)

Below left: D5901 and D5902, Welwyn North, 4 May 1969. Up Ballast. (Peter Foster)

Below right: D5901 and D5902, Welwyn North, 4 May 1969. (Peter Foster)

Operations: A High-Level Summary • 175

Above left: **D5902, Welwyn Garden City, 1 July 1969.** 08.43 King's Cross-Baldock passenger. (Peter Foster)

Above right: **D5903, Knebworth, 23 June 1966.** 2B65 King's Cross-Royston. (Peter Foster)

Below left: **D5900 and D1515, Hitchin South, 13 April 1965.** D5900 on 5B21 12.25 Cambridge to Ferme Park passing D1515 with 3N09 12.08 King's Cross to York parcels/sundries on 13 April 1965. (Chris Burton)

Below right: **D5909 and D9005, Hitchin, 31 August 1965.** D5909 awaiting departure with 2B64 16.36 service to King's Cross as D9005 passes on the afternoon 1A46 'Talisman' service for Edinburgh. (Chris Burton)

D5909, Hitchin, 26 September 1970. Relegation to an almost exclusive diet of freight and engineering work as a temporary substitute for Brush Type 2s in Doncaster Works for the fitting of air-brake equipment. (Courtney Haydon [RCTS Archive])

15.3 1970-1971: D5905 and D5909 Continue On

D5905/09 continued in traffic for a further two years after the withdrawal of D5900-4/6-8, with two reasons cited for their stay of execution:

- Deliberate retention for use on engineering trains in the Hitchin area, reputedly due to drivers' preference for the draught-free cabs of the 'Baby Deltics' compared with the Brush Type 2s, and/or,
- Temporary cover for Brush Type 2's out of traffic for the fitment of air-brake equipment.

Occasional use on passenger turns still occurred with the 08.43 King's Cross-Baldock and 16.31 Baldock-King's Cross services being the favourites. They were also apparently used to ferry train crews between Hitchin depot and various sites.

Chapter 16
DETAILS AND DIFFERENCES

16.1 As-Built
16.1.1 Front-End Arrangements
The photograph of D5905 below illustrates the main nose-end features.

The purpose of the nose-end access ladders was to facilitate cleaning of the cab windows by drivers, secondmen or fitters. Two grab handles were provided on the top of the bonnet to assist holding on. There were no nose-end bonnet top access doors as fitted to the English Electric Type 3 and Type 4 classes.

D5905, 30A Stratford, July 1963.
(Colin Whitfield [Rail-Photoprints])
Front-end detail:
1. Train identification discs (all in closed position); 2. Disc securing clips; 3. Tail lights; 4. Nose-end access ladder; 5. Communicating/gangway doors; 6. Gangway door securing clips; 7. Lamp brackets; 8. Fire alarm pull handle; 9. Warning horns (behind grille); 10. 'Blue-star' multiple control jumper socket; 11. 'Blue-star' multiple control jumper cable (with plug in picture located in the dummy socket); 12. Engine control air pipes (inner pipes); 13. 'Blue-Star' coupling code identifier; 14. Vacuum brake pipe, 15. Trip-cock equipment (system isolating cock); 16. Trip-cock equipment ('banjo' trip switch and associated vacuum pipework); 17. Steam heat pipe; 18. Air reservoir equalising pipes (outer pipes); 19. Drawgear (coupling shackle and hook); 20. Electrification warning flash; 21. Works Plate; 22. 34G depot plate.

D5905, Stratford DRS Yard, 26 June 1960. '2-1' side (No.1 end nearest). Key features:
1. Large radiator grille cover in position as fitted when new; the grille was removed as part of the impending workshop attention.
2. Train heating boiler air intake grille above nearest bodyside window, a '2-1' side feature only.
3. Battery box access doors on fuel and water tank sides, another '2-1' side feature only.
4. No bogie shock absorbers.
5. Two bogie sandboxes (two per bogie, one each side).
6. No shed plates fitted; bear in mind that D5905 had only been allocated to 34G Finsbury Park for two months at this stage.
(David Dunn Collection)

D5903 and D5902, Stratford DRS Yard, 17 August 1962. '1-2' side (No.1 end nearest). Key features:
1. Large radiator grille cover removed to improve airflow.
2. Train heating boiler access panel immediately behind the No.1 cab door (a feature of the '1-2' side only).
3. Boiler ventilation grille between access panel and large radiator grille (another '1-2' side only feature).
4. No battery box doors on fuel and water tanks on '1-2' side.
5. No bogie shock absorbers.
6. Two bogie sandboxes (two per bogie, one each side).
7. 34G shed plate below locomotive number at No.1 end. (David Dippie [David Dunn Collection])

D5904, approaching Hadley Wood South Tunnel, 19 October 1961. New Barnet station just out of view in the distance. Roof apertures (from No.1 end [nearest] to No.2 end): train-heating boiler aperture, large circular radiator fan grille, engine exhaust port. ((R.G. Warwick [David Dunn Collection])

No foot recesses were let into the locomotive body below the cab door kick plates as on the English Electric Type 3 and Type 4 locomotives. Instead, a step was welded immediately beneath the locomotive frame.

16.1.2 Bodyside arrangements
Bodyside details and changes are illustrated in the photographs on the previous two pages.

16.1.3 Roof
See photograph below.

16.1.4 Bogies
Sand boxes
It will be recalled that the 'Baby Deltics' were originally envisaged to carry eight bogie-mounted sandboxes but, as a weight saving issue, they were introduced with four in the outer positions on each bogie.

Further weight saving was undertaken in mid-1960 when the steel boxes were replaced by aluminium alloy, manufactured by English Electric and fitted at Hornsey. The weight of the aluminium alloy boxes was 58lbs per locomotive set (empty) compared with 176lb per set for the steel boxes.

Equalising beam and cross-member holes
All ten 'Baby Deltics' were delivered new with the equalising beam and cross-member holes drilled out.

Bogie shock absorbers
Shock absorbers between the equalising beams and the bogie frame were also due to be fitted as new but were subsequently omitted as one of the weight-saving initiatives. However, following introduction, the shock absorbers were fitted to most locomotives retrospectively at Stratford with only D5903 not fitted by the time of the class's removal from traffic in 1962/63 prior to rehabilitation. D5903 was released from rehabilitation with the equipment fitted.

16.1.5 Trip-Cock Equipment
Trip-cock equipment was fitted to the 'Baby Deltics' to enable safe operation over the London Transport Executive Metropolitan Widened Lines between King's Cross York Road and Moorgate in the up direction and Moorgate and King Cross Platform 16 (14 from 1972) in the down direction.

The equipment in simple terms included two key components:

- An isolating cock, located on the buffer beam on the 'Baby Deltics', which isolated the alternative trip-cock loop in the vacuum circuit during normal BR operations, but which brought the loop into operation when set by train crew.
- The all-important trip-cock, or trip-switch (colloquially called the 'banjo'), which was located immediately forward and slightly 'outside' of the leading wheels on the secondman's side of the locomotive. When rotated downwards into position by train crew this switch was

D5907, 30A Stratford, 28 April 1963. Two sandboxes per bogie; one each side. Drilled equalising beams. Bogie fitted with shock absorbers immediately adjacent to the helical springs. (Rail-Online)

D5905, 30A Stratford, July 1963. Trip-cock equipment; system isolation stop-cock 'switch' visible below the multiple-unit jumper cable socket, together with additional vacuum pipework incorporating the trip-cock 'banjo' and lever visible below the sandbox. (*Colin Whitfield [Rail-Photoprints]*)

capable of being tripped by lineside equipment in situations where the locomotive passed a red light. Tripping of the switch automatically destroyed the vacuum and resulted in an emergency brake application.

Various other classes received the trip cock equipment including selected members of Classes 15, 16, 21, 24, 26, 28 and 30/31. The method of installing the equipment varied between the Classes but the underlying principle was the same.

In reality, the use of the 'Baby Deltics' on the King's Cross to Moorgate route was patchy in the extreme; their tendency to emit acrid exhaust fumes in the tunnels between King's Cross and Moorgate resulted in a deliberate attempt to avoid rostering on these services.

182 • ENGLISH ELECTRIC TYPE 2 BO-BO 'BABY DELTIC' LOCOMOTIVES

D5906 and D5904, 30A Stratford, 7 April 1963. Trip-cock equipment on D5906's bogie; 'banjo' trip-switch crank mechanism and additional trip-cock vacuum hose circuit in evidence below the sand box although the tripping lever itself appears to have gone missing. (Frank Hornby [Colour-Rail])

16.1.6 Works Plates

D5905 English Electric/Vulcan Foundry works plate. EE Rotation No. followed by VF Works No. Note the 1958 date despite the weight rectification issues which delayed delivery until 1959. (Alan Whincup)

16.1.7 34G shed plates

Shed plates were fitted under the locomotive number at the No.1 end only on both sides (i.e. under the driver's window on '1-2' side and under the secondman's window on '2-1' side).

As far as is known, 34B Hornsey shed plates were never fitted. However, a photograph of D5902 on 23 April 1960 (see page 75) does illustrate cabside securing bolts but without plates; one day later all of the 'Baby Deltics' were re-allocated from 34B Hitchin to 34G Finsbury Park, so the bolts might reflect either the removal of 34B plates or preparation for the fitment of 34G plates.

The metal 34G shed plates were carried until 1968 when painted 34G digits were somewhat crudely substituted on those remaining in traffic.

16.1.8 Electrification Warning Flashes

From 1960s the 'Baby-Deltics' started to receive electrification warning flashes and for obvious safety reasons these were strategically positioned adjacent to the nose-end ladders and the alongside the bodyside boiler water filler doors. Four flashes were provided for each locomotive:

Nose-end flashes:
- Either, to the right of the gangway doors (GD) and ladders (broadly on a level with the top GD hinge although the precise position varied somewhat between locomotives): D5900-2/4/5/9
- Or, on the right-hand gangway door itself (roughly on a level with the top GD hinge although again the precise position varied between locomotives): D5903/7/8.
- Or, a combination arrangement, with the flash to the right of the GD and ladder on No.2 end and on the right-hand GD door on No.1 end: D5906.

Bodyside flashes:
- On the 1-2 side of the locomotive the flashes were generally positioned to the right of the boiler water filler door, whereas on the 2-1 side the flashes were positioned to the left side. The flashes on both sides were broadly on a level with the door handles although, as ever, exact heights varied.
- D5908 was an 'odd-ball' with the flashes significantly higher with the bottom of the flash at the same level as the top of the filler door, and, just to be awkward, the flash on the 1-2 side was positioned to the left of the door.

16.2 Post-Rehabilitation
16.2.1 Modified Front-end Arrangements

Following rehabilitation, the train reporting discs and associated clips were removed together with the gangway doors and access ladders. Four-character headcode boxes were fitted and the marker lights were repositioned at a lower level.

Brian Haresnape (*British Rail Fleet Survey*, Vol. 1): '… one definite improvement [of the refurbishment process] was the removal of the gangway doors and headcode discs from the nose ends, which were

D5909, Hitchin s.p., 18 August 1969. Sandboxes, works plates and 34G depot plates removed. (Peter Foster)

Front-end detail: 1. Four-character headcode box (operated from inside nose of loco); 2. Relocated tail lights; 3. Lamp brackets; 4. Warning horns (behind grille); 5. 'Blue-star' multiple control jumper socket; 6. 'Blue-star' multiple control jumper cable (with plug in picture located in the dummy socket); 7. 'Blue-Star' coupling code identifiers; 8. Vacuum brake pipe; 9. Steam heat pipe; 10. Coupling shackle and hook; 11. Electrification warning flash.

considerably neater in appearance with an integral four panel route indicator added.'

In my humble opinion, more finessed headcode boxes, as fitted to the EE Type 4s D345-D399 and the Type 5 'Deltics', would have been far superior as opposed to the very basic angular version actually fitted. That probably reflects fitment of the BR-designed boxes at a cost of £306 per locomotive, compared with the EE-designed boxes which would have cost £450.

16.2.2 Cab Draught Screens
See photograph and caption at the bottom left of this page.

16.2.3 Bogie Sandboxes
All locomotives were delivered post-rehabilitation with two sandboxes per bogie (one each side) fitted towards the outer ends; this arrangement was unchanged compared with their pre-refurbished condition. However, very soon thereafter these boxes were removed and based on photographic evidence D5900/4/5/8/9 had had the boxes removed by the end of July 1965. The other five very likely had them removed around and about the same time; photographic evidence certainly indicates that D5903 was devoid of boxes by January 1966 and the other four by July 1966.

16.2.4 Electrification warning flashes
Post-refurbishment the 'Baby Deltics' continued to carry four electrification warning flashes, one on each nose end and one on each bodyside. Following the major front-end design work it was inevitable that the position of the nose-end electrification flashes moved; however, the personnel at English Electric Vulcan Foundry managed to apply them uniformly across both ends of each of nine of the ten locomotives i.e. to the right of the newly fitted head code boxes (with 30 per cent of the flash height *above* the level of the top of the headcode box and 70 per cent *below* - see photograph on page 183். For some reason D5901, the last locomotive released from Vulcan Foundry, was different from the other ten, and, different at each end as well, as follows:

No.1 end: Top of flash 30 per cent of flash height <u>below</u> the level of the top of the headcode box.
No.2 end: Top of flash 10 per cent of flash height <u>above</u> the level of the top of the headcode box.

At some point, D5908 received an additional electrification warning flash on each front end, unique to this locomotive. The extra flashes were positioned to the left of the headcode boxes, with the 30/70 per cent 'rule' applied.

As far as the bodyside flashes were concerned (i.e. adjacent to the boiler water filler door on each side), it appears that in most cases the position of the flashes were left unchanged, suggesting that they were left *in situ* during the rehabilitation and repainting process. The one obvious exception was D5908 where the 'odd-ball' flash on the 1-2 side of the locomotive was moved to the 'standard' position to the right of the boiler water filler door with the bottom of the flash at the same level as the door handle; the flash on the 2-1 side was similarly positioned at the lower level.

16.3 D5901
D5901, whilst in Vulcan Foundry for refurbishment and fitment of 'U' series engine, received additional bodyside grilles (see page 129). These were subsequently removed when the project was cancelled and the locomotive converted back to standard.

Below left: D5909 and D5905, 30A Stratford, 3 January 1972. Cab draught screens, fitted as part of the rehabilitation process, are visible behind all cab seats. 34G depot codes crudely painted over the scars of the previously fitted metal plates. Different locomotive font styles are also clearly evident. (John Grey Turner)

Below right: D5905, 34G Finsbury Park, 11 July 1965. All sandboxes removed. (Keith Guthrie [RCTS Archive])

Chapter 17
LIVERIES

17.1 GNY

The assigned Appearance Design Consultant for the 'Baby Deltic' was Professor R.D. Russell, although it is evident from the literature that Ted Wilkes of Messrs. Wilkes and Ashmore was employed by BR at a later stage to cover the livery aspects.

English Electric were informed on 16 November 1956 that the number series for the 'Baby Deltics' would be D5900-9.

The BR Sponsoring Engineer at Doncaster corresponded with English Electric on 26 August 1958 regarding the painting and numbering of the Type 4 locomotives (D200) and EE used this as the basis for their Type 2s. In late August/early September EE asked BR if it was acceptable to proceed on this basis. The tone of the memos and letters at this stage suggest the lack of a nominated Appearance Design consultant to cover livery matters; Ted Wilkes was subsequently recruited to cover this gap, in addition to his other BR responsibilities.

By 16 September 1958, the BTC Design Officer (George Williams) was requesting EE for copies of their livery drawings. The drawings were forwarded on 19 September (see side elevation diagram opposite).

On 23 September George Williams (BTC) explicitly deployed Ted Wilkes to produce the official painting specification and colour layout for the 'Baby Deltics' 'in the usual form as early as possible'. Wilkes responded on 13 October 1958 with the following letter:

Locomotive Liveries: English Electric Type 2
Enclosed is a colour drawing showing our recommended colour scheme for the English Electric locomotive, as requested in your letter dated the 23rd of September. A colour specification is (*also*) attached.

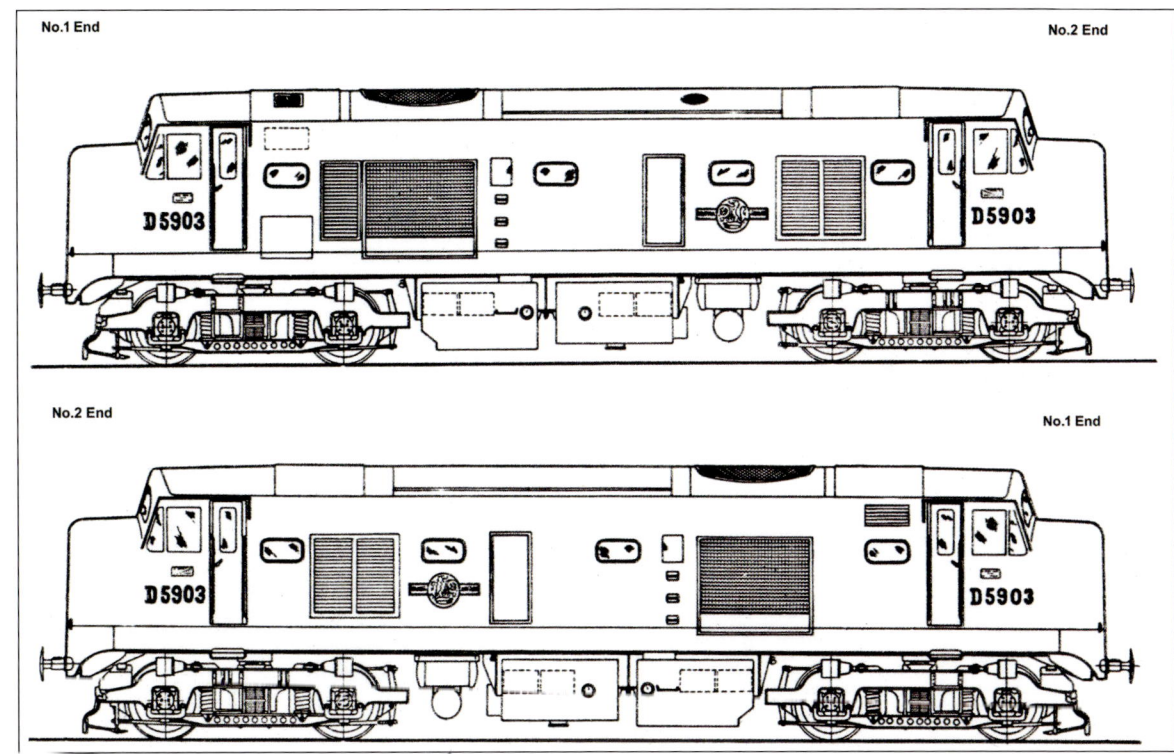

Styling Proposal for the 'Baby Deltics (based on EE Drawing No. P3295/006 dated 17 September 1958 held at the National Archives, Kew). Note the position of crests; these were subsequently moved at BR's request and located in a more central position beneath the second bodyside window in from No.1 end on both sides, rather than the third window). (BR Main Line Locomotive Layout Diagrams, *Undated*)

Notes on the colour scheme: I have recently been studying the appearance of the new BR locomotives of the past two years and have come to the conclusion that the application of linings is one of the least satisfactory features. In many cases there is no alternative but to employ a lining as the only means of enlivening an otherwise drab and poorly shaped locomotive, and such a lining seems to be most successful when it is in the traditional position - namely at waist level. But generally speaking I think that linings should be avoided if possible, even if only because they seem to have a faintly old fashioned look, just as with cars.

On this locomotive, we are suggesting that the lower part of the body be painted in a contrasting colour. The light grey that we have agreed for the Birmingham [BRC&W] locomotive seems to be the obvious choice if we are to preserve some sort of consistency.

However, partly in order to give added tonal contrast with the indicator discs and also because this locomotive has no visible buffer beam, we suggest that the continuity of the line be preserved but the colour changed to bright red around the ends of the locomotive.

In the absence of a visible buffer beam, I believe English Electric sometimes paint the front cross member of the bogie frame in bright red, but this will not need to be done in the case of the scheme as shown.

We do not recommend accentuating the gutter moulding in a contrasting colour because in this design the moulding is high up on the roof curve and cuts across the roof shape as it descends the front pillars. It is not such a good or important line as it appears in the side elevation drawing.

Note that the roof colour is stopped short of the actual end of the roof because the roof appears to end in a crease line or peak and it would look weak to end the colour on the peak.

We recommend that the body sides be devoid of any lining. The way in which the dark green continues over the roof curve as far as the gutter (as on some Southern Region carriages) can be quite attractive and gives something of the effect of a stressed skin fuselage. The isolated and evenly spaced windows with their rounded corners add to this fuselage effect. As soon as these windows are tied-up in any way by a lining, the effect is lost.

We also recommend a smaller crest than that shown on the English Electric print. We have used the smaller size (14¾" diam.) on other locomotives and it is perfectly satisfactory. We have also re-positioned the crest so that it does not appear quite so hemmed in by the door, a window, and louvre unit.

We have also re-positioned the locomotive number and shown it in Grot [font].

Incidentally, the front horn grille castings should be aluminium or body colour and not brass. You may remember that the other English Electric locomotives had brass grilles that looked very out of place.

General Instruction for Painting Main-Line Diesel-Electric Locomotives.
Type & Make of Locomotive: Type 2 by English Electric.

Items	Paint Specification
1. Body (upper).	Std. Loco Green BR Spec. 30.
Body (lower).	Light grey to BSS. 2660, colour 5-058, changing to Bright Red, colour 0-005 around ends.
2. Roof.	Mid Grey to BSS. 2660, colour 9-100.
3. Bogies.	The new underframe colour.
4. Undergear details.	The new underframe colour.
5. Buffer casings.	Bright Red to BSS. 2660, colour 0-005.
6. Drawhooks, guide, etc	The new underframe colour.
7. Lining.	None.
8. Numerals and prefix.	White.
9. Window frames.	Black rubber glazing strip.
10. Additional items.	*Nothing specified.*

Williams forwarded Wilkes' paint specification and colour layout proposal to R.C. Bond (BRB CME) on 20 October with a suitable cover note:

English Electric Co. Type 2 (1,100hp) Painting and Numbering
I have now received our Consultant's recommendation in regard to this locomotive, and enclose three copies of the painting specification, together with Wilkes' colour drawing.

You will appreciate that apart from the use of the standard green and roof colour, the details of the proposed livery differ slightly from other locomotives which have been finished to date. I assume, however, that we are not at this stage bound to standard livery treatment for all locomotives, and that we are permitted to treat each locomotive on its own merits. It is obvious, however, that some form of standardisation will be essential eventually, but I believe that carrying out minor experiments in lining and such details will help us to achieve a good and recognisable standard treatment in the future.

Briefly, Mr. Wilkes does not recommend bodyside lining at waist level for the above locomotive but suggests that the lower part of the body be painted in the contrasting grey colour. The standard grey used for the Birmingham locomotive has been selected to preserve a degree of consistency.

As this locomotive has no visible buffer beam, it is recommended that the continuity of the grey band be kept, but change to bright red around the ends of the locomotive., it being assumed that the red will still be required as a warning of approach. It is recommended that the smaller size (14¾in diameter) of crest used on other new locomotives should be used in this case in place of the larger version shown on the English Electric styling drawing.

Finally, Mr. Wilkes would recommend that the front horn grilles should be in aluminium rather than brass. At this stage of the building, however, it is realised that this is likely to be impossible, and we therefore recommend that the grille should be finished in the body colour.

The 'General Instruction for Painting Main-Line Diesel-Electric Locomotives' submitted by Williams to Bond was slightly modified by the following additional information or modifications:

Bond agreed to the proposals and he forwarded the relevant documents onto the Sponsoring Engineer at Doncaster for forwarding on to EE, as per the standard administrative procedure. Williams also informed Wilkes of the decision; regarding the underframe colour, Williams commented, 'We have asked the Chief Mechanical Engineer to try out this new colour on one of the Birmingham locomotives, and dare not specify it for all new locomotives until we have had the opportunity of considering it with him.'

English Electric were informed by the BTC that it would not be necessary for the buffer faces or wheel tyres to be painted. English Electric Drawing No. P.3202/059 was explicitly marked: "Area on top of nose to have matt finish to paint to stop reflection", presumably another BTC stipulation.

The somewhat unusual Grot font was utilised for the locomotive numbers with serif-D prefixes as recommended by Wilkes. The horn grilles were painted red and the 'new underframe colour' was not deployed.

Items	Paint Specification
3. Bogies.	Black
4. Underframe details.	Black
6. Drawhook guide, etc.	Black.
10. Top of outside footplating	Standard.
11. Builder's plate, size, position and colour.	Standard builder's plate. Position as shown on drawing.
12. Crest	Small crest, diameter 14¾".
13. Additional Miscellaneous Items.	*Nothing specified.*

Above left: D5909 and D5904, Stratford DRS (Works Yard), August 1962. GNY. A 'Grot' font was specified and the characteristics of the font seem to be more pronounced on some locomotives than others, probably the result of hand painting. Serif-D prefixes were applied to all locomotives. Both locomotives stored out of use. (C.R.G. Stuart [Colour-Rail])

Above right: D5900, King's Cross, May 1965. GSY. (F. Hornby [Colour-Rail])

Tom Grieves (BR Engineer) commented in his book *The Trials and the Triumph* (2012), 'When delivered, the "Baby Deltics" literally failed to shine – they arrived with matt paintwork and had to go back to English Electric for varnishing.'

A design critique was included in the September 1959 edition of *Trains Illustrated* covering 'The "Small Deltics" of the Eastern Region' as follows:

The last of the Type 2 diesels to appear in service, the English Electric 'Small Deltic', is a very interesting machine which differs from the other manufacturers' Type 2 designs… In contrast to some of the other diesel locomotive designs to appear so far a very striking effect has been achieved by the bold use of colour, with complete success. The superstructure is in dark olive green with a broad white band along the length of the body sides, at the bottom edge, connecting with a similar band of red across the front ends over the buffer beam (surely just as effective a warning to permanent way men as the ugly yellow "vees" still prominent on the front ends of diesel railcars in this country).

All of the 'Baby Deltics' retained the GNY livery at the point they were 'shopped' for rehabilitation.

17.2 GSY

Following rehabilitation, all ten 'Baby Deltics' emerged in two-tone green livery based on standard BR Locomotive Green, with a lighter Sherwood Green (BR Specification 30A, Item 58) base band around both the sides and the front-ends of the locomotive. Small yellow panels enveloping the new four-character headcode boxes were applied. The conventional black underframe and grey roof remained. The buffer beam was painted red. The 'Grot' font was replaced by the standard BR font with the Serif-D retained. The BR crest size and positioning remained unchanged.

The livery closely resembled the livery carried by the EE Type 5 'Deltics.

17.3 GFY

In compliance with lineside safety standard developments, several 'Baby Deltics' received full yellow nose-ends while still in green livery; the locomotives concerned were D5900/3/4/8, with D5908 also receiving the new British Rail corporate 'double-arrow' logo.

Interestingly, the February 1968 edition of the *Railway Observer* records D5909 as having received GFY livery. The *Classic Diesels & Electrics* magazine (No.8, November/December 1998) carried an article 'Green Dream' in which it was stated, 'D5909 was unique in that it carried all over yellow ends in both green and blue. It lost its green livery during a works visit in January 1968.'

No evidence has been found of D5909 in GFY livery and the *Railway Observer* report is believed to be erroneous, subsequently reiterated by CD&E.

GFY Repaint Dates - Relevant sighting information:

D5900   Doncaster Works: 23/08/68
D.L. Percival first GFY sighting: 07/09/68.

D5903   Doncaster Works: 05/03/67
D.L. Percival first GFY sighting: 14/03/67.

D5904   Doncaster Works: 05/02/67. Newark: 11/02/67 (en route to 34G from Doncaster Works)
D.L. Percival first GFY sighting: 14/02/67.

D5908   Doncaster Works: 01/10/67 and 15/10/67
D.L. Percival first GFY sighting: 30/11/67 (plus double-arrow emblem).

Notes:
1. The retention of the lion and wheel emblem by D5900 in September 1968 is somewhat surprising particularly given the application of the double-arrow emblem to D5908 nine months previously.
2. D5901/5/7 are known to have visited Doncaster after D5908 in October/November 1967 and it is perhaps surprising that these too did not receive full yellow ends and double-arrow emblems.

17.4 BFY
D5909 passed through Works in early 1968 and, following

D5903, Peterborough, 7 October 1968. GFY. (Bill Wright [Rail Image Collections])

D5908, 34G Finsbury Park, 8 June 1969. GFY. Double-arrow emblem in the same position used for the previous lion-and-wheel emblem. The emblem was smaller than those subsequently applied to D5909. Two electrification flashes on nose-end; unique to this locomotive (and applied to both ends). (R Holland [David Dunn Collection])

D5909, King's Cross, Undated. BFY, a retrograde step in my opinion; Monty Wells (*Railway Modeller*, May 1983) goes further suggesting that D5909 became "an ugly, pugnacious machine".
(Tom Bowman [courtesy "griffith_p"])

repairs, emerged repainted in all-over standard Rail Blue livery with full yellow ends, the only member of the Class so treated. Disproportionately large 'double-arrow' emblems were positioned below each cab side window. Rail Alphabet 'block' numbers and non-Serif D-prefixes were repositioned on the body sides. The roof was also repainted blue.

Given the recommendations of the 1967 National Traction Plan, it is surprising that D5909 was so treated; a touch-up and varnish GFY treatment would have been more appropriate.

BFY Repaint Date - Relevant sighting information:

D5909 Doncaster Works: 04/01/68, 07/01/68, 09/01/68, 17/02/68.
Ex-Works: 16/02/68 (DD29).
D.L. Percival first BFY sighting: 19/02/68

17.5 'D'-Prefixes
'D' prefixes were retained by all ten locomotives at withdrawal. This applied equally to D5901 after Derby RTC duties.

17.6 Data Panels
Position at withdrawal:

Data panels applied: D5900-2/5/9

Data panels not applied: D5903/4/6-8

The data panels on D5900-2/5 were applied under the driver's cabside windows on each side, beneath the numbers. The panels on blue-liveried D5909 were applied beneath the bodyside numbers at the driver's end on both sides.

Chapter 18
STORAGE AND WITHDRAWAL

18.1 National Traction Plan: 1965

The BR Re-shaping Plan (published 27 March 1963) specified extensive route rationalisation and withdrawal of passenger services across the whole of British Railways as well as new initiatives such as 'merry-go-round' operations for coal traffic and the significant development of container services. It was recognised that this Plan, plus the Freight and Coal Concentration and Freight Sundries Plans would significantly impact future motive-power requirements.

The need for detailed plans to fully understand forward locomotive requirements was identified and T.C. Baynton-Hughes was made responsible for the production of such plans. The first National Traction Plan (NTP) was published in February 1965. This involved a total re-appraisal of locomotive requirements on a national scale based on (i) revised future passenger and freight traffic requirements and (ii) acceptable standards of locomotive availability and utilisation. The Plan defined the numbers of locomotives required by Type for each operating Region, identifying any surpluses or shortfalls, and the supporting maintenance and servicing infrastructure required to achieve the required levels of locomotive availability.

At the end of 1964, the NTP identified the total stock of main-line diesels as 2,463 locomotives (including electro-diesels), increasing to 3,101 once all outstanding orders had been completed and delivered. This compared with an assessment of locomotives required at the end of 1963 of 3,731.

Future requirements were determined based around forecast train mileages, train speed improvements (15 per cent increase assumed), freight train payload increases, locomotive utilisation improvements (to 12 hours hauling trains per day), motive power availability improvements (87 per cent for Type 2 locomotives) and various contingency factors. The re-assessment of traffic requirements and performance improvements resulted in a re-defined locomotive requirement of 3,101 locomotives, albeit with identified surpluses in the Type 2 and Type 3 categories (28 and 77 respectively) and deficiencies in the Type 1 and Type 4 categories (57 and 48 respectively). Judicious national deployment of locomotives was seen as the solution to this issue.

Being cynical, one wonders whether the efficiency improvements deployed in the plan were deliberately set to achieve the neutral result between locomotives delivered and on order (3,101) and those needed to cover future requirements (3,101)! The result, however, was the avoidance of £55m capital investment (at 1965 prices) i.e. 630 fewer diesel locomotives.

The NTP identified the need for 975 Type 2 locomotives within the 3,101 total (based on the prescribed 87 per cent availability); this compared with 1003 actually in traffic or ordered. The Plan included an assessment of the impact of worse than Plan availability levels; thus at 85 per cent availability the requirement for Type 2s increased to 996, at 80 per cent 1070, and at 75 per cent 1132. The Plan indicated that the availability rate for Type 2s during the period April-December 1964 averaged 82.1 per cent, 5 per cent short of the NTP basis assumption. Subsequent problems in achieving the required 87 per cent level impacted on the timing of the phased elimination of steam on BR.

The NTP included the ten 'Baby Deltics' in the make-up of the nominally 975-strong Type 2 fleet, but referenced the fact that the 'Standard' Type 2 design was the BR/Sulzer diesel-electric with AEI electrical equipment. The recorded availability of the latter locomotives was given as 86 per cent.

The availability of the 'Baby Deltics' during 1964 was presented in Appendix F (Sheet 3) of the report, although the statistics were somewhat meaningless in that this was the year when these locomotives were being re-introduced to traffic following refurbishment (commencing July 1964). During the 13 weeks ended 16 December (at which point eight locomotives had re-entered traffic) availability averaged approximately 48-50 per cent. Availability then jumped to 70 and 80 per cent during week ending 23 and 30 December 1964 respectively. A subsequent 'one-off' report for week ending 13 April 1966 showed availability for the 'Baby Deltics' as 70 per cent, as compared with 84 and 87 per cent for the BR/Sulzer 1,250hp and the re-engined Brush/EE (1470hp) locomotives respectively.

18.2 National Traction Plans: 1967 and 1968

The NTP (November 1967) reviewed the period 1967-74 in the light of 'the predicted fall off in coal traffic, the effects of development of the freight grid, and expansion of freightliner and company trains, together with the rationalisation of both passenger and other freight services.' Plans were, therefore, drawn up to meet 'the immediate commercial load... year by year'. As a consequence, the overall BR requirement for locomotives in the Type 2 power category was forecast to reduce from 1002 (end-July 1967) to 724 (end-1974), and for the ER specifically from 367 to 210.

In terms of rationalising the fleet, the following factors were considered: locomotive performance and reliability, low availability over long periods, elimination of minority types, locomotives costly/difficult to maintain, locomotives known to be technically inferior, etc). The 'Baby Deltics' fell foul of many of these criteria and were dealt with accordingly in the forecast Type 2 composition; by 1967 the 'Baby Deltics' were very much considered to be 'minority' and 'non-standard'. The 'Baby Deltics' were massively numerically inferior to the BR (478), BRCW (116) and Brush (263) Type 2 classes and it was inevitable that the class would be eliminated early on. The unique nature of the Napier 'Deltic' T9 engine and relatively poor availability performance compared with other Type 2s added strength to the argument for disposal. The 1967 NTP anticipated the phasing-out of the 'Baby Deltics', together with the comparative fleet numbers of the other *diesel-electric* Type 2 classes, is given below:

1967 National Traction Plan.

Type	End-1967	End-1968	End-1969	End-1970	End-1971	End-1972	End-1973	End-1974
EE/Napier (1100hp)	10	10	10	4	1	-	-	-
MV/Crossley (1200hp)	20	10	-	-	-	-	-	-
NBL/MAN (1000hp)	8*	5	-	-	-	-	-	-
NBL/Paxman (1350hp)	20	20	18	8	8	8	-	-
BR/Sulzer (1160hp)	151	150	150	150	150	114	59	19
BR/Sulzer (1250hp)	327	327	327	327	327	327	327	327
BRCW/Sulzer (1160hp)	47	47	47	47	47	47	47	47
BRCW/Sulzer (1250hp)	68	68	68	68	68	68	68	68
Brush/EE (1470hp)	263	263	263	263	263	263	263	263

*Assumed 30 already withdrawn during 1967.

Notes:
1. During 1968 the requirements for Type 1s could be fully met by the Type 1 power currently available without recourse to any Type 2 substitution which had previously been the case.
2. With respect to the 'Baby Deltics', the NTP specifically recorded, 'The 10 English Electric 1,100hp locomotives are fitted with a 'Deltic' engine which has proved expensive to maintain. In the interests of rationalisation this minority type should be withdrawn.' Withdrawal by the end of 1971 was envisaged.
3. In terms of the other Type 2 classes, the 1967 NTP saw the Metrovicks and the NBL/MAN D6100s completely withdrawn during 1968, the 'Baby Deltics' during 1971 and the NBL/Paxman D6100s during 1972, leaving just the BR, BRCW and Brush Type 2 classes.

The NTP (December 1968) updated the 1967 Plan to reflect the latest commercial and economic position. The anticipated requirement for Type 2 locomotives was forecast to reduce to 635 by end-1974, with 203 of these allocated to the Eastern Region; see below:.

1968 National Traction Plan.

Type	Stock at 05/10/68	End-1968	End-1969	End-1970	End-1971	End-1972	End-1973	End-1974
EE/Napier (1100hp)	9	8	-	-	-	-	-	-
NBL/Paxman (1350hp)	20	20	10	-	-	-	-	-
BR/Sulzer (1160hp)	149	149	142	103	88	46	10	-
BR/Sulzer (1250hp)	327	327	327	327	327	327	327	327
BRCW/Sulzer (1160hp)	47	47	47	47	27	27	14	-
BRCW/Sulzer (1250hp)	68	68	68	68	68	68	68	65
Brush/EE (1470hp)	263	263	263	263	263	262	250	243

Notes:
1. Some adjustments to the availability and utilisation standards were applied in the 1968 Plan, as follows:
   Availability: Classes 25, 27 and 31 set at 87%, higher than for Classes 24 and 26 (80 and 75 per cent respectively).
   Utilisation: Increased to 12.5hrs hauling trains or shunting per day.
2. The 1968 Plan envisaged an earlier demise of the 'Baby Deltics' and NBL/Paxman Type 2s compared with the 1967 Plan.

However, evolving circumstances dictated further plan changes. A nationwide upturn in traffic levels during 1969, maintained through 1970, resulted in the deferral of the withdrawal and elimination of the remaining 'Baby Deltics', with final withdrawal ultimately taking place in March 1971.

18.3 Collateral Financial Implications of BR's Decision to Withdraw the 'Baby Deltics' from 1968

In May/June 1968, BR informed EE that they would not require any more repaired T9-29 engines for the 'Baby Deltics', apart from the two which were under repair at Napier's on 2 May 1968. At the point when withdrawal of the 'Baby Deltics' commenced, English Electric carried large stocks of spares for the T9 engines assuming a 'normal' life span for the locomotives and the Napier engines.

With the plan for Doncaster Works to take over engine overhaul from September 1968 BR had originally intimated that it would ultimately purchase the spares; however, at the point when D5906 was withdrawn, there was no written agreement to this effect. Brian Webb (1982) quantified the scale of the problem. 'Napier had provided staff, spares lists, tools, jigs, fixtures, etc, to enable Doncaster to take-over engine work at a cost of some £9,000 and had a commitment to the work said to be in the region of £150,000.'

Extensive negotiations were required to resolve this issue. The following BRB Supply Committee memorandum and meeting minutes shed some light on this difficult subject, made all the more complicated by amalgamating three areas of dispute:

Memorandum to the Supply Committee, 21 October 1970
Supply of Electrical Equipment by English Electric – A.E.I. Traction Ltd
1. We have been in dispute with EE-AEI Traction Ltd for some time concerning three major problems as indicated below:
2.1 Owing to late delivery by EE-AEI of electrical equipment for rolling stock for the Bournemouth Electrification Scheme the Board retained £16,000 as liquidated damages.....The Company [EE] required the

Board to demonstrate that the Company had failed in timely performance of the Contract, that the Board suffered a loss from a delay by the Contractor and in addition particulars of such portion of the goods as could not be used commercially in consequence of the delay. Nevertheless, they offered to settle, without prejudice at £2,000.

2.2 Following the Board's decision to discontinue the use of the 'Deltic' Type 2 locomotives which had been maintained by EE-AEI under contract, the Company sought to recover the cost of redundant spares and equipment remaining in their possession. These spares have no application to other types of locomotive, and are of scrap value only. A maintenance service could not have been provided without preserving an adequate stock of spares and although the Board has no contractual liability it does have some moral liability on this account. Termination of maintenance was without warning to EE-AEI and in any case as they had believed the work would be taken over by BREL they had not anticipated the locomotives would be scrapped. Also had the locomotives remained in service the Board would ultimately have required to purchase these spares for BREL to use in maintenance. The value of the redundant spares claimed by the Company was £138,000.

2.3 Under an agreement the 'Deltic' Type 5 locomotives had been maintained by EE-AEI. Following termination and handover of maintenance to BREL the CM&EE BRB considered certain pistons were defective and required to be replaced at a cost of £65,000. Accordingly a claim was made against EE-AEI on the grounds that during their contract they had supplied defective pistons. EE-AEI contended the pistons were within the tolerances permitted by the specification and refused to accept any liability.

3. It has not been possible to reach a settlement on individual cases and negotiations took place to clear all three simultaneously. Subject to the Board agreeing to the withdrawal of the claim for the Type 5 pistons the Company have agreed:
   - To reduce their claim for redundant Type 2 spares from £138,000 to £80,000.
   - Spares to become the property of the Board for which it is estimated that £2,000 will be realised as scrap.
   - The damages for late delivery of the Bournemouth stock being assessed at £8,000.

4. It is recommended that these three long outstanding disputes be settled on the above basis.

Signed: F.G. Manning, Supplies Department.

Supply Committee Minutes of Meeting, 5 November 1970
Minute 702. Supply of Electrical Equipment by English Electric – A.E.I. Traction Ltd
"The Committee considered a memorandum from the Supplies Manager, dated 21 October 1970, recommending a basis of settlement of outstanding disputes with English Electric - AEI Traction Ltd. in the late delivery of Bournemouth electrification stock, claim for redundant 'Deltic' Type 2 locomotive spares and claim for defective pistons for 'Deltic' Type 5 locomotives.

"The Committee considered that the proposed settlement of the claim in respect of the redundant 'Deltic' Type 2 locomotive spares for which there is no contractual liability should be reconsidered having regard to the cost incurred by the Board in connection with the defective pistons on 'Deltic' Type 5 locomotives."

How the dispute was ultimately resolved is not known.

18.4 Withdrawal Administration

To keep the financial records and asset registers straight and to ensure conformity with the prevailing NTPs, formal retrospective approvals for the condemnation of locomotives was requested for each calendar year.

Thus, a Memorandum dated 8 August 1969 requested formal authority for the condemnation of 421 locomotive withdrawn during 1968 i.e. 158 diesel main-line locomotives, 251 shunters and 12 electric locomotives, this including two 'Baby Deltics'. Relevant details were as follows:

Locomotives	No.	Original Cost	Written-Down Book Value	Residual/scrap Value
D5906/7	2	£158,100	£ 86,955	£4,300

Technically, D5900/3 were also withdrawn in 1968, but as the accountancy period was Period Ending (4 weeks) 25/01/69 these two locomotives were included in the 1969 withdrawals submission.

D6330 and fourteen Clayton Class 17s, withdrawn in 1968, were subsequently re-instated to traffic during 1969; in addition, two further Claytons were transferred to Departmental Stock. As a consequence, the final authority for condemnation, requested in January 1970, was for 404 locomotives.

A similar memo in August 1970 requested formal authority for the condemnation of 158 locomotives withdrawn in 1969, including six 'Baby Deltics':

Locomotives	No.	Original Cost	Written-Down Book Value	Residual/scrap Value
D5900-4/8	6	£474,295	£284,575	£ 11,979

Another Memorandum, this time dated August 1972, requested authority for the condemnation of 362 locomotives withdrawn in 1971, including D5905/9:

Locomotives	No.	Original Cost	Written-Down Book Value	Residual/scrap Value
D5905/9	2	£158,100	£ 82,212	£ 1,380

The average residual scrap value for the ten locomotives was £1766 each, although individual residual scrap prices varied considerably: D5900 £1879, D5901/2 £1900 each, D5903/4/6/7 £1,250 each, D5905/9 £690 each and D5908 £2000. The very low price for D5905/9 presumably reflected the absence of engines/generators for these two locomotives.

18.5 Storage/Withdrawal: Numerical Order

See Section 6.1 (specifically items highlighted in red).

18.6 Final Withdrawal: Chronological Order.

N.B. Excludes any reference to storage dates.

Loco. No.	Withdrawal Date	Notes
D5906	30/09/68	
D5907	20/10/68	
D5900/3	30/12/68	
D5904	20/01/69	
D5908	09/03/69	
D5902	23/11/69	
D5901	07/12/69	Transferred to Departmental Stock 1we 06/09/69
D5905	14/02/71	
D5909	07/03/71	

Notes:
1. D5906 was taken out of traffic in May 1968 pending a power unit change but, despite two power units being under repair at Napier's Works in early May 1968, the decision was taken to condemn the locomotive on 30 September 1968, with D5907 following on 20 October.
2. D5901, although officially withdrawn from BR stock on 7 December 1969, was initially loaned to Derby RTC on 31 August 1969, the loan becoming from permanent on withdrawal from revenue-earning stock.
3. D5902 is frequently quoted as being stored at Hitchin depot in January 1969 with defective tyres. However, the locomotive history for D5902 (see Section 8) shows extensive use of the locomotive during 1969, together with an Unclassified repair at Stratford DRS in June 1969. It was not until mid-November that D5902 suffered from the loose tyres issue, with withdrawal following on 23 November 1969.
4. D5905/9 survived for a further fifteen months, with D5905 being withdrawn on 14 February 1971 and D5909 on 7 March 1971.

18.7 Time in Traffic

Raw data:

Time in Traffic – 1.

Loco. No.	Ex-English Electric (B. Webb)	Date to Hornsey (BLS/CD&E)	Stopped Pre-Rehabilitation (*Deltic Deadline*)	Re-introduction to Traffic (B. Webb)	Date Withdrawn (RSL)
D5900	22/05/59	08/06/59	25/01/63	02/10/64	30/12/68
D5901	22/05/59	xx/xx/59	28/01/63	29/04/65	1we 06/09/69 to RCD 07/12/69
D5902	01/05/59	12/05/59	16/04/62	27/11/64	23/11/69
D5903	17/04/59	24/04/59	14/02/62	04/09/64	30/12/68
D5904	24/04/59	01/05/59	01/05/62	01/07/64	20/01/69
D5905	08/05/59	08/06/59	13/06/63	20/12/64	14/02/71
D5906	08/05/59	11/05/59	05/10/62	29/10/64	30/09/68
D5907	15/05/59	22/05/59	13/11/62	31/03/65	20/10/68
D5908	29/05/59	05/06/59	23/08/62	18/08/64	09/03/69
D5909	19/06/59	xx/xx/59	07/03/62	16/07/64	07/03/71

Notes:
1. The total elapsed time in traffic calculation in the table opposite is based on 'Date to Hornsey' to 'Date Withdrawn'
2. The time out of use pending and during refurbishment calculation is based on date 'Stopped Re-Rehabilitation' to 'Re-introduction to Traffic' dates.
3. Net time in traffic calculation is based on (1) minus (2).
4. Assumed 'Date to Hornsey' information to facilitate the calculations for D5901 and D5909 are 30/05/59 and 17/07/59 respectively.
5. Withdrawal date used for D5901 is 06/09/69 (i.e. date removed from revenue-earning traffic).

Calculated days in traffic:

Time in Traffic – 2.

Loco. No.	Total elapsed time in traffic (days)	Time pending/ undergoing re-habilitation (days)	Net time in traffic (days)	Time out of traffic as % of total elapsed time in traffic
D5900	3493	616	2877	17.6
D5901	3772	822	2950	21.8
D5902	3848	956	2892	24.8
D5903	3538	933	2605	26.4
D5904	3552	792	2760	22.3
D5905	4269	556	3713	13.0
D5906	3399	755	2644	22.3
D5907	3439	869	2570	25.3
D5908	3565	726	2839	20.4
D5909	4251	862	3389	20.3
Total	37212	7887	29325	21.2
Av. per Loco	3721	789	2933	21.2

Notes:
1. D5905 had the longest life of 11.7 years.
2. D5906 had the shortest life of just 9.3 years.
3. Allowing for the time out of traffic pending and undergoing refurbishment, D5905 had the longest operational life of 10.2 years, and D5907 had the shortest life of 7.0 years.

D5909 and D5905, 30A Stratford, 3 January 1972. Dereliction prevails! (John Grey Turner)

Chapter 19
STORAGE LOCATIONS

19.1 36A Doncaster and Doncaster Works

19.2 34D Hitchin

Above left: D5906, 36A Doncaster, June 1968. (Rail-Online)

Above right: D5907 and D5906, Doncaster Works (Erecting Shop Yard), 15 June 1969. (Author's Collection)

Left: D5905 and D5909, 34D Hitchin, 21 February 1971. (C. Loddington [Colour-Rail])

19.3 Finsbury Park

Above left: D5903 and D5904, 34G Finsbury Park, 18 January 1969. (Gordon Lacy [via 'Chronicles of Napier'])

Above right: D5900, D5903, D5908, D8233 and D5904, 34G Finsbury Park, 22 March 1969. (David Percival)

19.4 Ferme Park Down Sidings

D5903/00/04/08, Ferme Park Down Sidings, 9 April 1969. Assembled at the closed Ferme Park Yard to avoid cluttering Finsbury Park over the busy Easter period (4 to 7 April). The LCGB *Bulletin* (March 1969) reported closure of the yard with effect from 27 January 1969. 'Shed Master Archives' also records D5900/3/4/8 at 34B Hornsey on 20 April 1969. Some comedian has wound the headcode in D5903 to '1X01', the Royal Train! (Peter Foster)

19.5 34G Finsbury Park (again)

Above left: D5908, 34G Finsbury Park, 8 June 1969. (R. Holland [David Dunn Collection])

Above right: D5904, 34G Finsbury Park, 8 June 1969. (R. Holland [David Dunn Collection])

Below left: D5900, 34G Finsbury Park, 8 June 1969. (R. Holland [David Dunn Collection])

Below right: D5903, 34G Finsbury Park, 8 June 1969. (R. Holland [David Dunn Collection])

Above left: D5905 and D5909, 34G Finsbury Park, 6 March 1971. (Peter Foster)

Above right: D5909, 34G Finsbury Park, 18 March 1971. Withdrawn and awaiting its final transfer to 30A Stratford for storage. (Graham Hardinge)

19.6 30A Stratford

Above left: D5900, 30A Stratford, 16 February 1969. Diesel Repair Shop behind the photographer; 'C' maintenance shed roof visible in left background. Locomotive withdrawn. At Stratford depot after component removal at the DRS. (C. Campbell [David Dunn Collection])

Above right: D5902, 30A Stratford, 22 August 1970. Withdrawn. Alongside 'New' Shed. (Anthony Sayer)

202 • ENGLISH ELECTRIC TYPE 2 BO-BO 'BABY DELTIC' LOCOMOTIVES

Top left: D5902, 30A Stratford, August 1970. Alongside 'New' Shed. (Colour-Rail)

Top right: D5905, 30A Stratford, 10 April 1971. (John Turner, Birmingham [Rail Image Collections])

Above left: D5905, 30A Stratford, 28 August 1971. (Anthony Sayer)

Above right: D5909, 30A Stratford, 28 August 1971. (Anthony Sayer)

Right: D5905 and D5909, 30A Stratford, May 1973. Adjacent to No.1 Diesel Shed in readiness for departure to Kettering. (Rail-Photoprints)

Chapter 20
DISPOSAL

20.1 Disposal Summary
Summary of disposals:

George Cohen, Kettering: D5900/2-5/6/9, plus D5907 which is currently in the category of 'Disposal not proven' (see Section 20.2.5)).

John Cashmore, Great Bridge: D5908.

Derby RTC: D5901 (for subsequent Departmental history and final disposal see Section 21).

20.2 G. Cohen, Kettering
20.2.1 Sold for Scrap
Information reported in the RCTS *Railway Observer* 'Withdrawn Locomotives' section:

RO October 1969:
Sold for scrap: G. Cohen, Kettering: July 1969: D5900/3/4 (from Hornsey) (*from official source*).

RO April 1971:
Information request regarding disposal of D5906/7.

RO July 1971:
Sold for scrap: G. Cohen, Kettering: July 1969: D5906/7 (*source presumed to be from RCTS membership*).

Sales details for D5902/5/8/9 were never reported.

D5906 and D5907, Doncaster Belmont Yard, 9 August 1969.
(Gordon Lacy [via 'Chronicles of Napier'])

Left: D5903 and D5900 (plus D5904 out of picture), Kettering Yard, Undated.
(Rail-Online)

Far left: D5904/00/03, Kettering Cattle Dock, June/July 1969.
(Author's Collection)

20.2.2 In Transit to Kettering

'Bungus the Fogeyman' appended a comment to a photograph of D5903 on the 'flickr' web site as follows:

> I remember an evening spotting at Glendon South when D5903 along with D5900 and D5904 were delivered to Kettering North Yard. The consist, to use the modern idiom, was sat at the home signal waiting to come off the Corby line for ages… Eventually the convoy, headed by D3 *Skiddaw*, got the road to Kettering… D3 returned north not long after light engine.

20.2.3 G. Cohen, Kettering: Scrapyard Location

George Cohen's Kettering yard was situated on the site of the Cransley Iron Furnaces between Kettering and the village of Cransley on the Loddington branch; this now long disused branch diverged westwards from the Midland Main Line south of Kettering station.

In road terms, the scrapyard was located north-west of the A43 road from Northampton to Stamford between Broughton and Kettering (see map on page 206).

The scrapyard no longer exists and the site is now an industrial estate located immediately west of the A14/A43 interchange.

20.2.4 Sightings

Sightings of 'Baby Deltics' en route to and at G. Cohen, Kettering:

Location	Date	Loco. Nos. (and comments)
Doncaster Works	26/07/69	D5906/7
Doncaster Belmont Yard	09/08/69	D5906/7
Kettering Goods Yard	03/07/69	D5900/3/4 (+ D3444/79/88/91 waiting entry into Cohen's yard)
	20/07/69	D5903 (+ D3444/88/91 waiting entry into Cohen's yard)
	21/07/69	D5903
	14/08/69	D5906 (N.B. D5907 not listed)
	03/09/70	D5902 (transfer 30A to Kettering 02/09/70)
G. Cohen, Kettering	05/06/69	Nil.
	06/07/69	Nil.
	07/07/69	D5900/4
	16/07/69	D5900/4
	20/07/69	D5900/4
	07/08/69	D5900/3
	14/08/69	D5900 (practically cut-up)/3
	30/08/69	D5903 (N.B. D5900 not visible in space next to D5903)
	01/09/69	D5903, plus D5900 (engine only)
	07/09/69	D5903/6
	21/09/69	D5903/6
	12/10/69	D5903/6
	23/10/69	D5906
	26/10/69	D5906
	31/10/69	D5906
	24/11/69	D5906

Location	Date	Loco. Nos. (and comments)
	27/11/69	**D5906**
	11/12/69	**D5906**
	26/12/69	**D5906**
	14/01/70	**D5908** (N.B. Incorrect RO 05/70 report, actually D5906)
	27/02/70	Nil.
	22/04/70	Nil.
	17/05/70	Nil.
	16/06/70	Nil.
	21/06/70	Nil.
	23/09/70	**D5902**
	24/10/70	**D5902**
	14/11/70	**D5902**
	20/04/71	Nil.
	30/06/73	**D5905/9**
	15/07/73	**D5905/9**
	12/08/73	**D5905/9**
	06/09/73	**D5905/9**
	15/09/73	**D5905/9**
	02/02/74	**D5909**
	24/03/74	**D5909**
	05/07/74	**D5909** (part-cut but not in scrapping area)
	17/02/75	Nil.
	22/02/75	Nil.

Additional Notes:
1. Kettering Goods Yard sighting (locos in order listed – from Shed Master Archives):
   03/07/69: D5904, D5900, D3479, D5903, D3444, D3491, D3488 N.B. No D82xx/D84xx recorded.
2. G. Cohen, Kettering sightings (locos in order listed – from Shed Master Archives):
   07/09/69: D5903, D2176, D8405, D8213, D8407, D8223, D8227, D3488, D8400, D8402, D8202, D3491, D3444, D5906
   12/10/69: D2176, D5906, D5903, D8227, D8202, D8402, D8400, D3444, D3491, D3448 (*probably D3488*), 12138, D3629, 15101, 15102, (E)20002
   24/11/69: D8402, D2176, 15101, D5906, 15105
   23/09/70: D2176, D5902
3. G. Cohen, Kettering sightings (locos in order listed – from K.C.H. Fairey):
   10/07/69: D5904, D5900… D3479, D2176… D8236, D8401, D8405, D8213, D8223… D8407, D8227, D8400, D8402, D8202
   07/08/69: D3444, D5903, D5900… D2176… D8401, D8402, D8202, D8227, D8407
4. Positions of locomotives in G. Cohen's scrapyard (all as viewed from the bridleway) (based on photographic evidence):

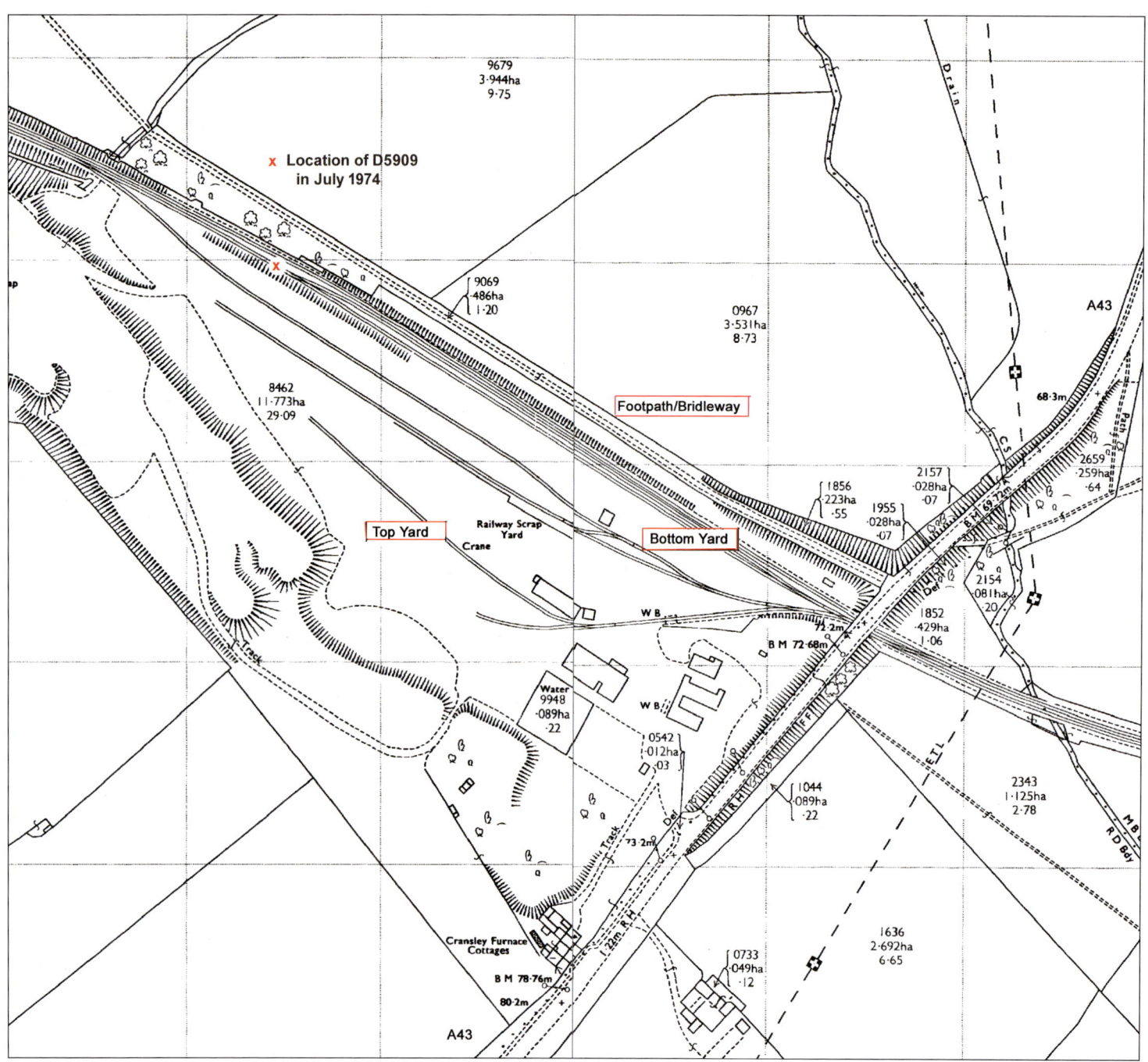

George Cohen's scrap yard, Cransley Sidings, west of Kettering. From Ordnance Survey 1:2500 map, 1972.

The A43 road and the "viewing" footpath/bridleway (north-east of the Kettering-Loddington line) are both marked on the map. The scrapyard could loosely be divided into two parts, the 'Bottom Yard' and the 'Top Yard'; these are my terms and were never the means of identification used by Cohen's themselves. The 'Bottom Yard' was composed of the two 'exchange sidings' immediately adjacent to the Kettering-Loddington line and the land immediately to the south of these sidings formed the main cutting area. The 'Top Yard' was the slightly elevated area to the south-west; this was the site of the former Cransley Iron Furnaces and associated internal railway system and in the Cohen era was largely used for storage of rolling stock awaiting scrapping. (© Crown Copyright and Land Information Group Ltd 2020 [Old-Maps.co.uk Ref.992903835]).

G. Cohen Kettering Plans.

10/07/69 (K.C.H. Fairey)

SE-NW

								1-2	2-1	1-2	2-1	1-2
								D8407	D8227	D8202	D8402	D8400
								Whole	Whole	Whole	Whole	Whole
	1-2	1-2		2-1	2-1	2-1	1-2	1-2				
	D5904	**D5900**		D8236	D8401	D8405	D8213	D8223				
	Whole	Whole		Whole	Whole	Whole	Whole	Whole				

07/08/69 (K.C.H. Fairey)

SE-NW	1-2	1-2			
	D5903	**D5900**	>>?		
	Whole	Part-cut			

14/08/69 (K.C.H. Fairey)

SE-NW

								1-2	2-1	1-2	2-1	1-2
								D8407	D8227	D8202	D8402	D8400
								Whole	Whole	Whole	Whole	Whole
	1-2	1-2			2-1	2-1	1-2	1-2				
	D5903	**D5900**			D8401	D8405	D8213	D8223				
	Whole	Part-cut			Whole	Whole	Whole	Whole				

30/08/69 (Author's Collection)

SE-NW		1-2			
		D5903	>>?		N.B. No evidence of D5900 to immediate NW of D5903.
		Whole			

21/09/69 (P. Foster)

SE-NW	1-2		1-2		
	D5906	Coach	**D5903**	>>?	
	Whole		Whole		

Above left: D5904 and D5900, G. Cohen, Kettering, 10 July 1969. Site of the former Cransley Iron Furnaces; by 1969 the furnaces had gone but a few service buildings remained. (K.C.H. Fairey [Colour-Rail])

Above right: D5903+D5900, G. Cohen, Kettering, 7 August 1969. (K.C.H. Fairey [(Colour-Rail])

Below left: D5900 G. Cohen, Kettering, 7 August 1969. (K.C.H. Fairey [Colour-Rail])

Below right: D5903 and D5900, G. Cohen, Kettering, 14 August 1969. (K.C.H. Fairey [Colour-Rail])

Disposal • 209

Above left: **D5903, G. Cohen, Kettering, August 1969.** After 14 August; see previous photograph. (Grahame Wareham)

Above right: **D5906 and D5903, G. Cohen, Kettering, 21 September 1969.** (Peter Foster)

Below left: **D5906, G. Cohen, Kettering, 21 September 1969.** (Peter Foster)

Below right: **D5906, G. Cohen, Kettering, Undated.** (Colour-Rail)

210 • ENGLISH ELECTRIC TYPE 2 BO-BO 'BABY DELTIC' LOCOMOTIVES

Above left: D5906, G. Cohen, Kettering, 11 December 1969. (John Evans)

Above right: D5902, G. Cohen, Kettering, 12 November 1970. (Nigel Petre Collection)

Below left: D5905 and D5909, G. Cohen, Kettering, 12 August 1973. (Fred Kerr)

Below right: D5909 and D5905, G. Cohen, Kettering, 12 August 1973. (Fred Kerr)

Left: D5909, G. Cohen, Kettering, 2 February 1974. D5905 has now disappeared from view but D5909 still lingers on. D5909 is repeatedly quoted in publications as being disposed of October 1973, but this is clearly not the case. (Fred Kerr)

Far left: D5909, G. Cohen, Kettering, 5 July 1974. Still extant and lurking around a corner amongst the vegetation, perhaps the reason for D5909's reputed earlier demise? The location of D5909 is marked on the map on page 206 for reference. Radiator elements removed. (Alistair Ness)

20.2.5 Disposal Discrepancies
Literature Search:
Details of 'Baby Deltics' cut-up at G. Cohen, Kettering are listed below using seven sources (both arrival and cut-up dates where provided); the comparators used are as follows:

- Strickland, D.C., *Locomotive Directory - Every Single One There Has Ever Been*, D&EG, 1983, plus relevant Supplements (Nos.3+4 and 7) (1983-87). (LocDir)
- Harris, R., *The Allocation History of B.R. Diesels & Electrics: Part 5 (Third & Final Edition)*, 2005. (AHBRD&E5)
- Marsden, C.J., *The Complete UK Modern Traction Locomotive Directory*, 2011. (MTLD)
- Butlin, A., *Diesel & Electric Locomotives for Scrap*, 2015. (D&ELfS).
- *Modern Locomotives Illustrated*, No.226, 'Class 23 and Class 28', August-September 2017. (MLI226)
- *Cohen's: A Northampton Railway Graveyard*, 2018. (Cohen's NRG) N.B. Author states dates supplied are cut-up dates but in reality they are more closely related to arrival dates.
- *MLIPlus*, No.259, pp20-29, 'The 'Baby Deltic' Story', February-March 2023. (MLI259)

Disposal Discrepancies - G. Cohen, Kettering.

Loco. No.	LocDir (inc Supplements 3, 4 & 7 (arr > cut-up)	AHBRD&E5 (cut-up)	MLTD (cut-up)	D&EfS (arr > cut-up)	MLI186 (cut-up)	Cohen's NRG (???)	MLI259 (cut-up)
D5900	07/69-08/69	08/69	06/69	06/69-07/69	06/69	06/69	08/69
D5902	09/70-10/70	10/70	08/70	09/70-09/70	08/70	08/70	08/70
D5903	07/69-10/69	10/69	06/69	06/69-10/69	06/69	06/69	06/69
D5904	07/69-08/69	08/69	07/69	06/69-07/69	07/69	06/69	07/69
D5905	06/73-10/73	10/73	08/73	06/73-10/73	08/73	08/73	08/73
D5906	07/69-11/69	12/69	07/69	08/69-07/71	07/69	07/69	07/69
D5907	07/69-xx/xx	08/69	07/69	08/69-08/69	07/69	07/69	07/69
D5908	01/70-02/70 (GCK)	01/70	-	-	-	-	-
	12/69-02/70 (JCGB)						
D5909	05/73-10/73	10/73	08/73	06/73-10/73	08/73	08/73	08/73

Comments:
1. On the basis of sightings below, the table has been colour coded as follows: correct information, incorrect information and, based on the absence of sightings and photographic information, unproven information.
2. D5908 was NOT cut-up at G. Cohen, Kettering (see Section 20.3).
3. Comparing sighting information with published information:

D5900	First seen 10/07/69; last seen 14/08/69 (part cut); not seen 30/08/69. C/U 08/69
D5902	First seen 23/09/70; last seen 14/11/70; not seen 20/04/71. C/U Disposal date not proven.
D5903	First seen 07/08/69; last seen 12/10/69; not seen 23/10/69. C/U 10/69
D5904	First seen 10/07/69; last seen 20/07/69; not seen 07/08/69. C/U 07-08/69
D5905	First seen 30/06/73; last seen 06/09/73; not seen 02/02/74. C/U Disposal date not proven.
D5906	First seen 07/09/69; last seen 14/01/70 (reported as D5908); not seen 27/02/70. C/U 01-02/70.
D5907	No sightings at GCK; the last sightings of D5907 were 09/08/69 at Doncaster Belmont Yard and at Chesterfield on an undated August day (both with D5906). At this point the plot gets interesting.

David Percival, a well known expert on all things 'Baby Deltic', records D5907's disposal as 'C, GB(?)' (i.e. John Cashmore, Great Bridge, but with a question mark) in his excellent 'End of the "Baby Deltics"' article in *Railway World* (May 1971). If I am honest, I would have rejected this possibility out of hand, but given Percival's known penchant for accuracy, this suggestion could explain D5907's apparent absence at Kettering.

Percival's subsequent equally outstanding articles in *Motive Power Monthly* (November 1991) and *Traction* (February 2007) made no further reference to D5907's disposal. Fast forward to the November 2019 *Today's Railways UK* article on the 'Baby Deltics' by Chris Booth and D5907 is listed as having been scrapped at J. Cashmore, Great Bridge in July 1969. In the following month's magazine, David Percival responds with the following comment on the Mail Train page:

'I think it is generally accepted that only D5908 was cut up by Cashmore at Great Bridge. There was certainly some doubt about the fate of D5907 when I wrote an article about the class in another magazine way back in 1971 but it was soon established that D5907 had been cut up at Cohen's yard in Kettering.'

Quite how it was 'soon established' is now unfortunately lost in the mists of time.
For me, the jury is still out. Disposal location and date not proven.

D5909	First seen 30/06/73; last seen 05/07/74 (part-cut but not in Scrapping Area). No subsequent visit reports; however, a photograph of D5909 (back in the Scrapping area and with cutting more advanced) did appear in Ashley Butlin's book *Diesels & Electrics for Scrap: Volume 1* (Volume 2 was never published). The photograph caption read: 'All, except D5901 and D5908, finally found their way to George Cohen's Kettering yard where in September 1973 demolition of D5909 was well under way…..only the cabs and bogies remained to be cut.' Should 1973 have read 1974, perhaps? Disposal date not proven.

20.3 J. Cashmore, Great Bridge
20.3.1 Sold for Scrap (Source: RCTS Diesel Dilemmas and RCTS member Andy Wylie)
D5908 was put up for sale, by BRB Supplies at Derby from Finsbury Park, on Sales Tender No.17/230/521T/103, with a copy of the tender sent to the CM&EE King's Cross on 18 August 1969 implying the tender was sent out about 11 August. Tenders were required by 8 September 1969.

The official date of sale of D5908 was 10 September 1969 and the CM&EE at King's Cross was advised on 12 September 1969 that the locomotive had been 'purchased for scrap by John Cashmore Ltd, Great Bridge, Tipton, Staffordshire'.

20.3.2 In Transit to Great Bridge
Jim Carter provided the following information with the three photographs illustrated opposite:

D5908 at Bescot in October, 1969. The first two slides show the locomotive on the 'Up and Down Goods' line (believed to be on 3rd October) whilst en route to John Cashmore's Scrapyard at Great Bridge for disposal. The first shot shows it was being hauled by a Class 47 locomotive. The second shot is a close up view of D5908 which, unfortunately, is not a 'pin sharp' image but still records a never to be repeated moment. What happened next is apparently a bit of a mystery! I understood the locomotive was moved to the scrapyard the day the first two images were taken but it was then found that the fuel tank was full and that the locomotive had to be brought back to Bescot for this to be drained. Another story indicates that the locomotive went direct to Bescot TMD due to a brake problem. The third view shows the locomotive, for whatever reason, later in the month on Bescot TMD. It finally moved to Great Bridge for scrapping in early January, 1970.

20.3.3 John Cashmore, Great Bridge: Scrapyard Location

The scrapyard was located about half a mile north-east of the old Great Bridge station on the east side of the now closed Stourbridge to Walsall line.

20.3.4 D5908 Sightings

G. Cohen, Kettering: 14/01/70 (*RO*, May 1970). N.B. Error: sighting believed to be D5906.

J. Cashmore, Great Bridge: 08/02/70 (*Rail Enthusiast*, September 1984).

Noted being cut-up at J. Cashmore, Great Bridge: xx/02/70 (*RO*, June 1970).

Over the years since late 1969, debates have raged over the disposal of D5908, specifically the disposal location at either G. Cohen, Kettering or J. Cashmore, Great Bridge.

Peter Hall has performed a considerable amount of research on D5908 and he has published the result of his deliberations in his excellent and highly informative RCTS *Diesel Dilemmas* web pages. It is worth reproducing his page concerning D5908 here:

Scrapping of 'Baby Deltic' D5908
If one locomotive disposal typifies the problem of repeating what has been published before it is that of 'Baby Deltic' D5908. Before explaining the saga in detail it should be said that it has now been established beyond doubt that D5908 was scrapped at J. Cashmore, Great Bridge in February 1970 and not at G. Cohen, Kettering.

Consultation of 1969/1970 ROs gives several sightings of D5908 at Bescot in late 1969, the final one being on 23rd November 1969 (p12, 1/70 RO). Previously it had been located at Finsbury Park. A further sighting at Bescot on 14th December 1969 is later quoted by Ashley Butlin. Roger Harris quotes in Part Five of *The Allocation History of BR Diesels & Electrics* – third edition (p59) that it was towed away from Bescot on 4th January 1970. It is thus fair to say that all references up until this date are correct.

The next published reference to D5908 is in the RO (p156, 5/70 RO) where it is quoted as being at G. Cohen, Kettering on 14th January 1970 along with D2176. Interestingly this report does not quote D5906 as present, which as far as can

D5908, Bescot Yard (Up and Down Goods Line), October 1969. Believed to be either 3 or 6 October. (Jim Carter)

D5908, Bescot Yard (Up and Down Goods Line), October 1969. (Jim Carter)

D5908, 2F Bescot, October 1969. (Jim Carter)

be ascertained arrived there in the autumn of 1969 and was not scrapped until July 1971 (*sic - see above*). Quite obviously D5908 was a typo for D5906 as two references contradict D5908 being at Kettering. Firstly, the RO (p.193, 6/70 RO) gives D5908 (from Finsbury Park) as noted as being cut up at J. Cashmore, Great Bridge in February. Secondly, when the subject was debated in *Rail* magazine in 1984 a reader wrote in stating that he had observed D5908 at. J. Cashmore, Great Bridge on 8th February 1970. Other than the 14th January 1970 reference to D5908 at Kettering no other observation has been traced of it en route from Bescot to Kettering or in the G. Cohen yard. Indeed, its presence at Kettering was queried in the RO (p193, 6/70 RO) but no resolution appears to ever have appeared in print.

What until now had not been published is information discovered by RCTS member Andy Wylie. Andy has viewed the DME KX records concerning D5908 and after condemnation on 9th March 1969 it was put up for sale, from Finsbury Park, on tender 17/230/521T/103. Unusually, it was a single item tender – most locos were sold in batches. BRB Supplies at Derby sent KX a copy on 18th August 1969 and the tender was due back on 8th September 1969.

The official date of sale was 10th September 1969 and KX were advised on 12th September 1969 that the purchaser was 'John Cashmore Ltd, Great Bridge, Tipton, Staffordshire'. Destination was Great Bridge and it was to be sent there a.s.a.p.

The copy of the Advice of Dispatch note (BR 8658) says that it was dispatched from 34G on 3rd October 1969.

A letter from the CM&EE York to KX on the 29th December 1969 states that "Messrs John Cashmore Ltd inform me that the above mentioned locomotive has not yet arrived at destination" and asks that KX trace forward.

Andy says that what happened next is not on the disposals file (KM5/160/6) but he has spoken to former DME KX colleagues of his who were there at the time; they traced D5908 forward to Bescot where the locomotive had been stopped 'with brake problems'. They said that it was consigned forward 'within days of us finding it'! This obviously taking place on 4th January 1970.

In addition to this, the 'HSBT Project', who are researching steam locomotive disposals, have discovered a reference in the J. Cashmore records showing the date taken in for D5908 as 5th January 1970.

With this fresh evidence now available it does appear beyond doubt that D5908 remained at Bescot until 4th January 1970 when it was towed to J. Cashmore, Great Bridge where it remained for several weeks before being scrapped sometime that February.

Pronounced Life Extinct, *A Decade After* and *Locomotive Directory* all quote D5908 as scrapped at G. Cohen, Kettering, presumably the former basing this on the erroneous Kettering sighting and the later two reproducing that error. Later titles, particularly after the subject was aired in *Rail*, tend to correctly record the disposal at J. Cashmore, Great Bridge. A problem though is that despite what had been published in the *Railway Observer* and *Rail* several quote a cut date of December 1969 when February 1970 is clearly correct. Interestingly though, despite what has now been published, Roger Harris still favours the G. Cohen, Kettering scenario in Part Five and Part Six (A) of *The Allocation History of BR Diesels & Electrics* – third edition (p59 & p112 respectively)!

The articles referenced by Peter Hall in his web page provide some additional background to the D5908 debate:

AHBRDE5, R. Harris, April 2005:

By 3/10/69 it had been towed away to Bescot, where it arrived on 6/10/69 and was stabled in the yard en route to J. Cashmore, Great Bridge, W. Midlands for scrap. However, all was not right, as the loco remained at Bescot, and on 23/11/69 it was stabled at the Diesel Depot, where it stayed until 5/1/70 when it was towed, this time to G. Cohen, Kettering, and noted intact in the yard on 14/1/70, but was then immediately cut up, the deed being done on 17/1/70.

AHBRDE6A, R. Harris, August 2006:

Many other sources quote this loco as broken up at John Cashmore, Great Bridge in 12/69.

(Note: However, the date of towing from Bescot was recorded as 5/1/70, so how it could be cut in 12/69 is a mystery!! It is said it was mistaken for D5906 at Cohen's, but as that loco was cut in 12/69, we still have the date of towing to contend with. Because it was due to go to Cashmore's does not mean it definitely went there.)

Rail Enthusiast, **September 1984,** p25, *The Scrapping of D5908* (Letter from A.K. Butlin):

With reference to the caption against the photograph of D5908 on page 35 of the July *Rail Enthusiast* regarding the disposal of this loco, readers may be interested to know that in-depth research concerning the disposal of D5908 reveals that it is highly unlikely that it… ever found its way to Cohen's of Kettering.

Withdrawn in 1969, it was towed during September or October of that year to Bescot where it was noted early in October at the back of the depot. Here it remained until at least the 14th December, when it appears to have moved the short distance to Cashmore's at Great Bridge, where its remains were noted during a visit in February 1970. Throughout the period December 1969-February 1970 weekly visits to Cohen's at Kettering revealed no sign of 5908, only 5906. It was D5906 which was wrongly reported as D5908 in a journal published during 1970, which led to the idea that D5908 made this strange journey to Cohens'.

All the evidence points to D5908 being cut by Cashmore's at Great Bridge during December 1969- February 1970. Perhaps a *Rail Enthusiast* reader has a photograph of D5908 during this period which would settle the question."

Rail Enthusiast, **September 1984,** p25, *'Baby Deltic' remembered* (Letter from: M. Harwood):

I too recall seeing 'Baby Deltic' D5908 with Clayton D8572 at Bescot in 1969 (*The Napier Sound*, July issue) but I feel I must put you right on the eventual last resting place of D5908. It was at Cashmore's, Wednesbury, that I observed D5908 on February 8 1970, in company with Class 11 shunters Nos. 12057/86 as I recall, and I can only assume that this is where D5908 met its untimely end. "

Interestingly, page 26 of the same magazine included a letter from Roger Harris entitled 'Eradicating errors' in which he criticises the author of an earlier *Rail Enthusiast* article (not related to D5908) for 'widespread mistakes', suggesting that it was 'the worst researched article I have ever read!' There is a certain irony here as it was Harris who perpetuated the RO error in AHBRDE5 and AHBRDE6A!

20.3.5 D5907?

David Percival (*Railway World*, May 1971) makes reference to the possibility that D5907 was disposed of in Cashmore's Yard, Great Bridge (see Section 20.2.5):

D5908… was eventually towed from Finsbury Park on October 3 to Bescot, where it remained for some time before being broken up at Cashmore's, Great Bridge. I believe that D5907 also ended its days there at about the same time.

D5907, with D5906, left Doncaster Belmont Yard on 11 August 1969, and if it did move to Great Bridge would have arrived there within a few days.

Cohen's at Kettering were typically not very fast at dismantling locomotives; in contrast, John Cashmore at Great Bridge disposed of its locomotives very rapidly and disposal via this route is theoretically possible. However, detailed perusal of the Cashmore's company entry register by Roger Butcher (leader of the HSBT *What Really Happened to Steam* Project) found no reference to D5907.

Solid sighting or photographic evidence is still required to prove that D5907 was indeed cut up at Kettering.

Chapter 21
DEPARTMENTAL SERVICE

During its early days, the Derby Railway Technical Centre relied upon available locomotives and crews for the operation of its test trains, a situation which resulted in numerous cancellations as a consequence of revenue-earning passenger and freight services taking priority. To resolve this unsatisfactory situation, the RTC looked to acquire its own locomotives. Metrovick D5705 was its first acquisition in late 1968, followed by D8512/21 in July 1969.

D5901 was transferred (in the administrative sense) from Finsbury Park to the Railway Technical Centre at Derby during week ending 6 September 1969 (31/08/69) initially on loan. No permanent re-allocation was published, and it assumed that this effectively took place when D5901 was officially withdrawn from BR stock during week ending 13 December 1969 (07/12/69). The first known physical sighting of D5901 at Derby was on 12 September.

D5901 was used by Derby RTC on various tests, including:

- 'Manual Speed Datum' tests (equivalent to modern-day 'cruise control') during 1970 on the Mickleover test track.
- the 'Auto Wagon' project during 1971 which involved the automatic loading and unloading of containers by terminal cranes and the automatic shunting of 'powered' container wagons by track to train transducers. In reality the wagons were unpowered with movements on the Mickleover test track performed by D5901).
- operation of the Tribometer or Tribology train on the main line from 1972 which necessitated D5901 visiting numerous

Below left: **D5901, Derby RTC, 28 September 1969.** (Peter Foster)

Below right: **D5901, South of Bedford, circa 1972.** Tribometer test train. (John Law)

Departmental Service • 217

locations which had never previously seen a 'Baby Deltic'.

The Tribometer train was used to assess wheel/rail interaction in both natural adhesion conditions and conditions post-treatment (e.g. water-cannon). The permanently coupled train consisted of a Laboratory Coach incorporating a control room, a four-wheel long-wheelbase COV-AB van and a driving-trailer coach. The COV-AB wagon housed the instrumentation and hydraulic packs, with special brake actuator units above each axle as well as tanks for laying fluid during the experiments; it had no conventional brakes and was fitted with end gangways for access.

The whole Tribometer consist was passed for 90mph running, although whether D5901 was given dispensation to exceed its 75mph limit is not known. It is believed that the train was never used with the driving trailer leading even though D5901 was fitted with 'Blue-Star' multiple-unit equipment.

The Tribometer train was used extensively during the 1970s with the train operated on a 180 mile circuit from Derby RTC to Birmingham New Street, Wolverhampton High Level, Stafford, Crewe, Stoke, Trent, Beeston and return to Derby, with intervening work on other routes suffering difficult adhesion conditions.

During service with Derby RTC D5901 suffered from the usual problems of radiator leakages and unburnt fuel and oil carry-over into the exhaust system. Etches Park dealt with the former as best it could (presumably with assistance

Above left: D5901, Sheffield, 21 June 1974. Tribometer test train. (Colour-Rail)

Above right: D5901 and 03397, Derby Etches Park, 29 November 1974. Into the TOPS era but D5901 continues on un-renumbered. (Stephen Dowle)

D5901, 36A Doncaster, 7 March 1976. Withdrawn. (Anthony Sayer)

D5901, Doncaster Works (adjacent to Scrapping Area), 13 June 1976. Still with the test train T21 reporting code rolled-up in the 4-character head-code box; photographic evidence indicates the prevalence of codes in the 1T20 to 1T25 range. Keeping company with 2173, 03091, and 24071. Both 24071 (as D5071) and D5901's successor at Derby RTC (24061, formerly D5061), together with D5901 itself, had all been fellow stalwarts on GN suburban operations in the early-1960s. (Anthony Sayer)

from Marston Excelsior), and the exhaust was largely solved by the simple expedient of avoiding long periods of idling.

D5901 remained in use until late 1975 when it was withdrawn from service. Steve Allsop in his article *Derby & the RTC* in *Traction* (September 1995) explained that withdrawal was due to an engine governor/load regulator fault which although relatively easily resolved by a visit to the Toton or Derby Works load bank facility. However, this was not to be. The Toton load-bank was unavailable for use at the time of D5901's problems and Derby Works was considered too expensive.

D5901 was booked to be transferred from Derby to Doncaster Works on 18 February 1976, but was noted at 41A Tinsley on 24 February and at 36A Doncaster on 29 February and 7 March 1976. Arrival at Doncaster Works had been achieved by 28 March. D5901 was cut-up between 27 January and 6 February 1977. As a direct consequence of its Research duties, it had become the last survivor of the 'Baby Deltic' fleet.

D5901 was numbered as such for all of its Research working life retaining the GSY livery throughout, surviving nearly twelve years without a repaint.

Tribometer operations from Derby RTC after D5901's departure were taken on by BR/Sulzer Class 24 24061 (subsequently RDB968007), which was withdrawn surplus to requirements during week ending 16 August 1975 (10/08/75), being first noted in the Derby Etches Park area on 6 September 1975. The sudden abandonment of many serviceable Class 24s allowed Derby RTC to replacing their faulty 'Baby Deltic' with something a little more conventional and RDB968007 proved to be a very reliable performer.

Readers are recommended to read Peter Hall's *Diesel Dilemmas* web page on the RCTS web site and Steve Allsop's *Traction* articles for further information regarding the 'Manual Speed Datum' project, the 'Auto Wagon' trials and the Derby RTC generally.

Chapter 22
PRESERVATION

22.1 D5901

An article entitled 'Preservation Pioneer' in the *Classic Diesels & Electrics* magazine (Issue 5, May/June 1998) included an interview with Chris Reid who, with Colin Massingham, acquired D821 for preservation in 1973. The interview included the comment: 'We also [subsequently] tried to go for "Baby Deltic" D5901 at one time and D6122 at Barry but that was too much money for us at the time.'

Maybe that explains why D5901 lingered for so long at Doncaster Works before finally being cut up.

22.2 D5910

D5910 is currently being 'built' by 'The Baby Deltic Project' at Barrow Hill Roundhouse. On completion D5910 will be made available for use on heritage lines; certification to main line standards is not anticipated. The new locomotive is being assembled using:

- Napier 'Deltic' T9-29 engine No. 388. This engine was originally extracted from D5905 as a spare for potential use in D5901 although it was never required. It was claimed by the National Railway Museum in March 1977 and stored at York from August 1977. The engine was subsequently purchased by a group of enthusiasts in 2001 and transferred to Colchester for further storage before moving on to Barrow Hill in September 2003, with the engine re-started in October 2008 following extensive refurbishment.
- A Class 37 body shell. The body shell of 37372 was acquired from the Harry Needle Railroad Company in February 2009, initially for use as an engine testbed. Subsequently, however, the body shell was 'promoted' to become the basis for new-build D5910 suitably modified to achieve 'Baby Deltic' proportions, and capable of accepting the Napier engine and associated equipment. A demonstration start-up of the 'Deltic' engine in 37372 was undertaken in August 2009.
- A pair of Class 20 bogies.

Details of the work involved in re-creating D5910 is fully described in The Baby Deltic Project's excellent book *Baby Deltic – The Story of an Engine's Re-Birth* (2nd edition, 2023), a story of total dedication and concerted engineering focus.

D5910, Barrow Hill (Roundhouse), 27 August 2022. As can be seen the body shell of D5910 has been substantially re-engineered to transform it from a Class 37 into an accurate recreation of an as-built 'Baby Deltic', involving shortening of the middle section between the cabs and both nose-ends. The cab window frames have been reshaped to 'Baby Deltic' proportions. The newly-refurbished and slotted Class 20 bogies are now indistinguishable from 'Baby Deltic' bogies and it remains to be seen how many sand boxes are to be fitted. (Anthony Sayer)

Chapter 23
CONCLUDING REMARKS

Many criticisms are directed at the English Electric 'Baby Deltics'. The use of the sophisticated 'Deltic' engine in a Type 2 locomotive has frequently been questioned and seen as an inordinately complex and costly solution to a basic requirement. Equally the use of such a complex engine at the dawn of wholesale dieselisation was seen as perhaps asking too much of the people and the prevailing infrastructure.

The Type 2 diesel locomotives designs of British Railways were all adversely impacted in one way or another by the 18ton axle-loading limitation and English Electric looked to solve the problem by the use of the lightweight high-speed 'Deltic' engine. That this innovative solution was ultimately thwarted by designers taking their eye off the weight of other key items of equipment was distinctly unfortunate.

Apart from the 'Deltic' engine, D5900-9 were fairly conventional, with the exception of the auxiliary gearbox arrangement and mechanically driven compressor. Only substitution of this compressor with a conventional electrically driven alternative during rehabilitation cured the gearbox issues.

There is no doubt that the 'Baby Deltics' suffered a significant number of engine and auxiliary equipment failures in their early years, necessitating numerous periods out of traffic for engine changes. English Electric did, however, point out that most of the locomotives' loss of availability was due to short periods of time spent in Works for engine changes and that, *when running*, the locomotives returned the best availability figures and the lowest maintenance costs of any BR Type 2 locomotive. A high level of availability was achieved by the 'repair by replacement' policy practised at Stratford, made possible by the lightweight 'Deltic' engines; so whilst the Works visits were frequent, they were usually fairly brief.

Perhaps the most telling statement was made by Tom Grieves (2012) concerning the 'unfortunate' combination of some of the mechanical problems suffered by the 'Baby Deltics' (gearbox and associated compressor drive shaft failures, together with consequential collateral damage to the coolant circuits) and the services which they performed (and he should know, he was living the job); I deliberately repeat Grieves' comments here:

[The 'Baby Deltics' had] an Achilles heel in that the compressor was manually driven through a drive shaft and clutch plate, the only such application within the new contracts. Their general reliability was far better than the other Type 2s but when the drive to the compressor failed it became a dead locomotive and could not limp home, so 'Baby Deltics' ran to time or not all. The fact that they were on higher profile services and had the most operation over two-track sections (so that when something went wrong they stopped the job) raised the profile to such an extent that they became widely criticised.

Many of these issues were never fully resolved until rehabilitation work was undertaken in 1963-65, and even then, oil leakages and radiator issues persisted. I still wonder why the Marston radiators were not substituted by the more reliable Spiral Tubes radiators, equipment which was extensively used in over 700 other English Electric locomotives.

The general consensus was that the rehabilitated 'Baby Deltics' were as good as any other Type 2s on the same Great Northern suburban passenger duties. The continued use of the two-stroke Napier 'Deltic' engine after rehabilitation did, of

Concluding Remarks • 221

course, mean that the characteristic blue exhaust problem persisted.

The 'Baby Deltic' fleet was also seen as unsuccessful when compared with the company's other products, notably the Type 1 and Type 3 fleets which have consistently enjoyed high levels of reliability and availability. This was indeed true due to the inherent ruggedness of the English Electric SVT/CSVT engines and the associated electrical equipment. The need for a succession of modification programmes to the Napier engines and the whole rehabilitation process severely marked the cards of the 'Baby Deltics', a situation from which they never seemed to fully recover.

The 'Baby Deltics' were also accused of being only the faintest shadow of their charismatic Type 5 cousins. This observation, whilst perhaps true in an absolute sense, was ultimately ridiculous and irrelevant. Type 2 duties were intrinsically different from Type 5 duties! However, we continue to see such strange comments as epitomised by a comment in an article in *The Railway Magazine* (August 2004, 'Baby Deltics' and Brush Type 2s'): 'The "Baby Deltics" were styled similarly to the "Deltics" themselves and, with their high-speed engines, the continuous drone on full power was a pale imitation of mighty Type 5 music. It was as if minions were attempting to win grandeur by donning the emperor's clothes and imitating his voice!'

Whilst operationally the engine horsepower of the 'Baby Deltics' at 1,100 was 265hp less than that of the Brush/Mirrlees Type 2s which they worked alongside for so long on GN suburban services, this was partly compensated for by the EE Bo-Bo's 32ton weight advantage, roughly equivalent to a saving of one coach. Hitchin drivers always seemed to be happy with their allocated motive power.

D5908 and D5900, Location unknown, 13 May 1968. (Colour-Rail)

The perceived and indeed real complexities of the Napier engines in a traction context in the late 1950s led to non-proliferation of the 'Baby Deltics' beyond the original ten 'Pilot Scheme' locomotives. Colonel H.C.B. Rogers, in his book *Transition from Steam* (1980) commented, 'J.F. Harrison [BTC CM&EE] did not think the ['Deltic'] engine [was] really suitable for railway traction because there were too many component parts.'

Indeed, A. Joyce (*Modern Railways* (September 1966) later commented:

> ...the 'Deltic' engine is a triangular, opposed-piston two-stroke running at 1500rpm, which is enough to damn any engine for UK rail traction use. Worse, it costs twice as much per installed hp as its 850rpm four-stroke cousin and needs overhaul every 4000-6000hr rather than a top overhaul at 15,000hr.

Following the 'Pilot Scheme' orders, J.F. Harrison's views ensured that it was the BR/Sulzer, BRCW/Sulzer and Brush/Mirrlees Type 2 designs which were proliferated (expanding to fleet sizes of 478, 116 and 263 locomotives respectively).

However, A. Joyce, again (*Modern Railways*, (October 1967), did comment:

> ... on the subject of service life it is worth mentioning... that between overhauls 'Deltic' maintenance is limited to changing injectors at 2000hr intervals. 'No valves, no tappets, no cylinder head joints, no rocker box covers,' as one engineer with 'Deltic' and orthodox engines in his care put it to me. It is true that although a medium-speed four-stroke will go to 10,000 or more before needing a top overhaul there are many minor and not-so-minor jobs to be carried out in that time.

Harrison's preference for the conventional low-speed engine, at least in a Type 2 context, effectively sealed the ultimate fate of the 'Baby Deltics' and, years later in 1967 with BR, via the National Traction Plan, looking to rationalise the diesel fleet, it was inevitable that the 'non-standard' and numerically-challenged classes would be axed first. And so it proved to be with the 'Baby Deltics'.

As with previous volumes, I use this opportunity to invite readers to assist with the following requests for information and assistance:

- BTC Driver's Instructions BR 33003/77 for D5900-9 specifically for line speeds of the three stages of traction motor field weakening.
- Diesel Locomotive Record Cards for D5901/2/5/9.
- Special Traffic Notices covering 'Baby Deltic' movements to scrapyards.
- Photographs, as follows:
  - At Vulcan Foundry under construction.
  - Undergoing repair *inside* Stratford DRS.
  - In transit to Vulcan Foundry for rehabilitation during July/August 1963.
  - D5900/2-9 at Vulcan Foundry undergoing rehabilitation.
  - D5909 at Doncaster Works in February 1968 in BFY livery.
- Sightings/photographs as follows:
  - In transit to Cohen, Kettering.
  - D5907 at Cohen, Kettering (or, alternate scrap yard).
  - D5901 being cut-up at Doncaster Works (late January/early February 1977).

SOURCES & REFERENCES

Books
Butlin, A., *Diesels & Electrics for Scrap: Volume 1*, Atlantic Transport Publishers, 1988.
Butlin, A., *Diesel & Electric Locomotives for Scrap*, Oxford Publishing Company, 2015.
Clough, D.N., *Diesel Pioneers*, Ian Allan, 2005.
Derrick, K., *Diesels & Electrics to the Scrapyard 1959-1989*, Strathwood, 2018.
Grayer, J., *Cohen's: A Northamptonshire Railway Graveyard*, Crécy Publishing, 2018, specifically Chapter 14, *Spotlight on the 'Baby Deltics'*, pp88-93
Greaves, T., *The Trials and the Triumph: A B.R. motive power engineer's experience of the steam to diesel years*, Bellcode Books, 2012.
Haresnape, B., *British Rail Fleet Survey: 1. Early Prototype & Pilot Scheme Diesel-Electrics*, Ian Allan, 1981.
Harris, R., *The Allocation History of B.R. Diesels & Electrics:* Third & Final Edition, *Part 5*, 2005 and *Part 6A*, 2006. (AHBRDE5 and AHBRDE6A)
Marsden, C.J., *The Complete UK Modern Traction Locomotive Directory*, TheRailwayCentre.com, 2011.
Percival, D., *King's Cross Lineside 1958-84*, Ian Allan, 1984.
Petre, N.A. and Carr, K., *The 'Deltic' Family 1955-1973*, Visions International, 2019.
Rogers, H.C.B., *Transition from Steam*, Ian Allan, 1980.
Stephens, R., *Diesel Pioneers – The British Rail Diesel Loco Fleet up to 1970*, Atlantic Transport Publishers, 1988.
Strickland, D.C., *Locomotive Directory*, D&EG, 1983, plus Supplements Nos.1-7 (1983-87). (LocDir)
Tufnell, R.M., *The Diesel Impact on British Rail*, Mechanical Engineering Publications, 1979.
Tuffnell, R.M., *Deltics Super Profile*, Haynes Publishing Group, 1985.
Webb, B., *English Electric Main Line Diesel Locomotives of British Rail*, David & Charles, 1976.
Webb, B., *The Deltic locomotives of British Rail*, David & Charles, 1982.
Williams, A., *BR Diesel Locomotive Album*, Ian Allan, 1979 (pp28-29).

The Baby Deltic Project, *'Baby Deltic': The Story of an Engine's Rebirth*, 2009 (1st edition) and 2023 (2nd edition). (BDP)

Professional Papers
C.D. Carmichael, *Design and Development of the 'Deltic' Engine*, Paper to Manchester Association of Engineers, November 1956.

Magazines
British Railways Illustrated, Summer Special No.7, Irwell Press, 1998, specifically:
   'Diesel Dawn: Unwanted Offspring', pp52-59.
Diesel Railway Traction, specifically:
   December 1955, pp363-370, *The Napier 'Deltic' Engine*.
   August 1959, pp301-302, *Small Deltic Locomotives*.
Trains Illustrated/Modern Railways (TI/MR), specifically:
   September 1959, pp434-435, *The 'Small Deltics' of the Eastern Region*.
   July 1960, pp421-423, *New Eastern Region diesel depot at Finsbury Park*.
   April 1962, pp266-270, *The Acceptance Testing of Diesel Locomotives*, R.K. Evans.
   May 1962, pp316-319, *The King's Cross suburban diesel scheme*, B. Perren.
   September 1966, pp505-506, *Traction: 'Deltic' Type 2s Redeemed*.
   June 1969, pp299-301, *Traffic Divisions of British Rail No. 12: King's Cross, Part 1*, B. Perren.
   July 1969, pp358-362, *Traffic Divisions of British Rail No. 12: King's Cross, Part 2*, B. Perren.
Railway World (RW), specifically:
   March 1969, pp102-108, *King's Cross today*, B. Perren.
   November 1970, pp513-515, *The Ashburton Grove Pullman*, P.R. Foster.
   May 1971, pp204-209, *End of the 'Baby Deltics'*, D.L. Percival.

October 1971, p460, Letters: 'Baby Deltics', P.A. Hogarth.
The Railway Magazine (RM), specifically:
August 2004, pp36-40, *Practice & Performance: Baby Deltics and Brush Type 2s*, K. Farr.
June 1973, pp273-275, *Mobile Adhesion Testing*.
Traction, specifically:
No.11, September 1995, pp4-8, *Derby & the R.T.C. Part 2*, S. Allsopp.
No.41, March 1998, pp18-21, *Baby Love*, J. Hypher.
No.147, February 2007, pp10-15, *Bye-Bye 'Baby Deltic'*, D.L. Percival.
No.150, April 2007, pp52-53, *'Baby Deltic' De-Railed*, D. Briars.
No.153, July 2007, pp36-37, *Pilot Scheme Locos*, S. Allsop.
No.219, March 2014, pp46-49, *'Baby Deltics'- Part 1*, S.C. Carter.
No.220, April 2014, pp18-21, *'Baby Deltics'- Part 2*, S.C. Carter.
Motive Power Monthly, specifically:
November 1991, pp30-35, *Class by Class: The '23s'*, D.L. Percival.
Today's Railways UK, specifically:
November 2019, pp54-59, *The Class 23 'Baby Deltics'*, C. Booth.
December 2019, pp56, *Mail Train: 'Baby Deltic' points*, D.L. Percival.
Modern Locomotives Illustrated, No.226, August-September 2017, *Class 23 'Baby Deltic' & Class 28 'Co-Bo' Fleets*, Key Publishing. (MLI226)
MLIPlus, No.259, February-March 2023, pp20-29, 'The "Baby Deltic" Story', Key Publishing. (MLI259)
Classic Diesel & Electrics (CD&Ex), specifically:
No.6, July/August 1998, pp40-45, *Perfect Ten*, D. Jarvis.
No.6, July/August 1998, pp52-60, *Take the Test! (Part 1, 1957-60)*, A. Bonson.

No.8, November/December 1998, pp18-22, *Green Dream*, M. Alden.
No.8, November/December 1998, pp58-61, *Final Test (Part 3, 1963-67)*, A. Bonson.
No.23, June 2008, pp24-27, *Red Alert!*, P. Mulhearn.
Rail Enthusiast, specifically:
February/March 1982, pp28-30, *Vulcan's Bad Bad Baby*, E. Bellass.
April 1982, p26, *Rail Mail: 'Baby Deltics'*.
July 1984, p35, *The Napier Sound on the GN Main Line*.
September 1984, p25, *Rail Mail: The Scrapping of D5908*, A.K. Butlin, and, *'Baby Deltics' Remembered*, M. Harwood.
Railway Modeller, specifically:
May 1983, pp184-189 and June 1983, p235, *'Baby Deltics'*, M. Wells.

Societies
Railway Locomotives, British Locomotive Society.
Deltic Deadline, Deltic Preservation Society (DD), specifically:
Issue 29, October 1982, 'How the Baby Became Blue', A.J. Wylie.
Issue 33, June 1983, '16th June 1963: D5905 Stored Unserviceable', A.J. Wylie.
Link, Engine Shed Society (ESS).
Journal, Great Eastern Railway Society, specifically:
October 2022, 'Tales from the Diesel Repair Shop: The Deltics', A. D. Nugent.
Journal, Ipswich Transport Society.
Bulletin, Locomotive Club of Great Britain (LCGB).
Railway Observer, Railway Correspondence & Travel Society, specifically:
June 1971, pp189-190, 'The "Baby Deltics"'.
Journal, Stephenson Locomotive Society.

Aurora, Stevenage Locomotive Society.

Official BR Sources
BR *Diesel Traction: Manual for Enginemen*, BTC, 1962.
BTC/BRB Works & Equipment Committee, Supply Committee, Technical Committee Minutes and Supporting Papers.
Diesel Locomotive Record Cards (BR.9215/1).
BR(ER) Statement of Diesel Casualties (Mechanical/Electrical) (BR 33582/2).
BR(ER) King's Cross District Locomotive Works Records (1959-64).
Stratford DRS Foreman/Chargehand Reports (A.D. Nugent and A. Rayment).

Websites
Chronicles of Napier.
Facebook: *34B 34C 34D Hornsey//Hatfield//Hitchin Loco*. Group.
RCTS *Diesel Dilemmas* (research by Peter Hall).
RMWeb, specifically:
On this Day in History.
RTC Main-Line Locomotives 1968-1978
Scrapping of 'Baby Deltic' D5908
Shed Bash UK.
Testing Times (The Tribometer Train).
The 'Baby Deltic' Project.
The History of First Generation DMUs.

Sightings
Commercial Sources: Shed Master Archives.
Individuals: A.J.Booth, R.Barnes, C Campbell, C.R.Capewell, D.Coddington, J.Collins, K.C.H.Fairey, G.Hardinge, M. Hunt, I.Osborne, D.L.Percival, A.P.Sayer, T.Skinner, J.Stretton.